T0222305

Springer Undergraduate Mathematics Series

More information about this series at http://www.springer.com/series/3423

Ieke Moerdijk • Jaap van Oosten

Sets, Models and Proofs

 Springer

Ieke Moerdijk
Department of Mathematics
Utrecht University
Utrecht, The Netherlands

Jaap van Oosten
Department of Mathematics
Utrecht University
Utrecht, The Netherlands

ISSN 1615-2085   ISSN 2197-4144   (electronic)
Springer Undergraduate Mathematics Series
ISBN 978-3-319-92413-7   ISBN 978-3-319-92414-4   (eBook)
https://doi.org/10.1007/978-3-319-92414-4

Library of Congress Control Number: 2018948806

Mathematics Subject Classification (2010): 03-01, 03-B10, 03-C07, 03-E25, 03E10

This Springer imprint is published by the registered company Springer Nature Switzerland AG
The registered company address is: Gewerbestrasse 11, 6330 Cham, Switzerland

Preface

This is the text of a one-semester lecture course on mathematical logic for second or third year mathematics students, which we have been giving at Utrecht University for about 20 years by now. The course is entitled *Foundations of Mathematics*. Actually, it is preferred that the students take the course in their third year, so they have more knowledge of what it is that they are studying the foundations of. In particular, they should have seen basic analysis, group theory and ring theory, as well as vector spaces more general than $\mathbb{R}^n$.

Students at Utrecht University and most other universities are subjected in their first year to a course that teaches them to write mathematical proofs; this course usually includes basic set-theoretic notation and elementary logic (truth tables for propositional logic, for example), which obviates the necessity for treating propositional logic in our course. Instead, we have opted to take the shortest route to the main first results from mathematical logic that have some mathematical significance, with a minimum of syntactical fuss. We believe this approach distinguishes our book from many other introductions to the field.

The format of our course was of course dictated by the university schedule of one-semester courses of 14 two-hour lectures. The main consequence of this was that we had to postpone several important topics, notably computability (recursion) theory, to a second more advanced course in logic. In the Appendix, we have tried to help the reader by giving some directions and suggestions for further study, both on the subjects treated in this book—model theory, proof trees and set theory— and in areas we did not treat such as computability theory and the incompleteness theorems.

While the global content of the course has remained fairly stable over the years, we have continuously made small improvements to the text while taking turns in lecturing the material. In particular, we have gradually added more and more exercises, of which there are now 146. These vary a lot in difficulty, ranging from easy "getting used to the notions" finger-exercises to serious exam questions. (In the latter case, fairly generous hints have been provided.) Sometimes, the result of an exercise is needed for the development of the theory, so one should do these in any case. Solutions to the problems have not been included. This is a matter of pedagogical principle, although we are aware that, in these times, solutions might easily appear on the web.

We would like to thank the teachers who have used our notes, some of whom have suggested to Springer to publish the notes as a book. We acknowledge the criticism and interest of all our students over the years. In particular, the following students and colleagues have given useful criticism, have suggested alternative proofs or exercises or have brought typos and other inaccuracies to our attention: Benno van den Berg, Eric Faber, Nicola Gambino, Jeroen Goudsmit, Joep Horbach, Peter Houben, Tom de Jong, Mark Kamsma, Fabio Pasquali, Marcel de Reus, Sebastiaan Terwijn, Andreas Weiermann, Ittay Weiss and Jetze Zoethout. Needless to say, all remaining errors are our responsibility.

We also thank the anonymous referees, who made valuable comments.

Finally, we acknowledge pleasant collaboration with Springer's editorial staff, in particular with Rémi Lodh.

Utrecht, The Netherlands   Ieke Moerdijk
                                                                                Jaap van Oosten

Introduction

Hilbert and the Motivation for Logic

In the year 1900, David Hilbert (1862–1943), one of the leading mathematicians of his time, addressed the International Congress of Mathematicians in Paris with a list of 23 problems, to be attacked in the coming century. For us, four of these problems are of particular relevance:

1. Settle the Continuum Hypothesis. This is the statement that every infinite subset of $\mathbb{R}$ is either in bijective correspondence with $\mathbb{R}$ itself or with $\mathbb{N}$. Is it true or false?
2. Prove that the axioms of the arithmetic of real numbers are free of contradiction.
10. Find an algorithm to determine whether a polynomial equation with integer coefficients has a solution in the integers.
17. Prove that a positive definite rational function is a sum of squares.

Mathematics had become, during the 1800s, more and more an axiomatic science. As Hilbert put it in his paper [33]:

> When we are engaged in investigating the foundations of a science, we must set up a system of axioms which contains an exact and complete description of the relations subsisting between the elementary ideas of that science.

Whenever we try to give such a system of axioms for a particular field of mathematics—be it group theory, Euclidean geometry or the theory of measure and integration—some natural questions arise:

1. Are the axioms *compatible*: don't they contradict each other?
2. Are the axioms independent, or can one axiom be derived from the others?

David Hilbert

3. Is the system of axioms strong enough? That is, they are supposed to govern a
   particular part of the mathematical world. Is every statement which is always true
   in this domain a consequence of the axioms?

Of these issues, the first one was of paramount importance for Hilbert. Indeed, it
was his philosophical conviction that *as soon as an axiom system is compatible, it
describes a part of mathematical reality.*

However, how does one *prove* that a certain collection of axioms is free of
contradiction? What does it actually *mean* to be "free of contradiction"? Similar
problems arise with respect to other questions. What does it mean to "derive"
something? Obviously, in order to study these matters in a precise way, we need
a mathematical apparatus to deal with "axioms", "compatibility" and "proofs".
Mathematical Logic is an area of mathematics in which such an apparatus is
developed.

Hilbert later formulated a more sophisticated motivation for setting up a formal
theory of "proofs". In his beautiful paper *Über das Unendliche* (On the Infinite,
[34]), he distinguishes two mathematical worlds: an *actual* world, directly accessible to inspection by the mind, the world of the integers and their elementary
properties and the geometry of Euclidean space; and an *ideal* world, where lots of
things live which have nicer properties than the actual things and whose description

is often more elegant. Often, we arrive at knowledge about the actual world via a detour through the ideal world. The examples Hilbert gives are:

- Imaginary numbers. Philosophers may doubt their existence, but we enjoy the fact that every polynomial has a complete factorization and the beauty of complex integration by which we also establish facts about the actual world (e.g. $\sum_{n=1}^{\infty} \frac{1}{n^2} = \frac{\pi^2}{6}$).
- Infinitesimals. Weierstraß's theory of limits and convergence shows how these can be reduced to finite numbers.
- Fractional ideals of number rings. Many rings, like $\mathbb{Z}[\sqrt{-5}]$, lack the desirable property of unique prime factorization, but this is restored if one turns to factorization of fractional ideals; this was shown by Kummer.
- And, could the world of infinite sets not also be seen as part of the ideal world?

The last item of this list was particularly dear to Hilbert, who calls Cantor's theory of sets "the most admirable flower of the mathematical mind, and one of the highest achievements of purely intellectual human activity whatsoever" [34]. The use of set theory and the higher cardinalities should be protected from paradoxes and be rendered absolutely uncontroversial. So, here is what Hilbert proposed:

Let us set up a formalism S in which we can formulate and prove results from this *actual* world. At the very least, such a system should be able to reason about natural numbers and prove their basic properties. Now the ideal world is inaccessible to S, but *proofs about* this world are finite things and therefore susceptible to analysis in S. Hilbert advocated the creation of *proof theory* (*Beweistheorie*).

What can we hope to achieve by studying proofs? Here, in our view, we can distinguish two forms of Hilbert's Programme: a weak and a strong version (weak and strong in aspirations and scope).

Hilbert's Programme, weak version: using the system S, which should be unproblematic in any philosophy of mathematics, prove that no proofs about the ideal (infinitary) world exist which have an absurd conclusion.

There is, however, evidence (e.g. in [34, 44]) that Hilbert envisaged stronger results, viz. that one could *eliminate* the use of infinitary reasoning. This led to:

Hilbert's Programme, strong version: suppose we are given a proof of some statement about the actual world (a statement which can be expressed in S), a proof which may have used the whole set-theoretic machinery. Then one should be able to find a (probably much longer and less elegant) proof in S of the same fact.

Please bear in mind that the "Programme" was never formulated very crisply and what we have given here is an interpretation. Discussion about what exactly Hilbert strove to do is still active to this day (see e.g. the philosophical paper [16]). In the Appendix to this book (Section A.2), we shall say a bit more on the fate of Hilbert's Programme.

What Is to Be Found in This Book?

We present an introduction to mathematical logic.

In Chapter 1 ("Sets"), we develop the mathematical theory of sets (the "paradise Cantor created for us", in the words of Hilbert) as used in everyday mathematical practice. The notions of "set" and "function" are taken as primitive. We recall elementary constructions of sets and functions and establish notation. Then we revisit cardinalities. Without having to say what "the cardinality of a set" actually *is*, we can say what it means that "the cardinality of the set X is less than (or equal to) the cardinality of the set Y": $|X| < |Y|$ (or $|X| = |Y|$). We prove the Schröder–Cantor–Bernstein theorem which says that if there are 1–1 functions $X \to Y$ and $Y \to X$, then the sets X and Y are in bijective correspondence. We show by Cantor's diagonal argument that every set X has lower cardinality than its powerset. We formulate the Axiom of Choice (AC) and use it to deduce that for every infinite set X, there is a 1–1 function $\mathbb{N} \to X$ as well as the familiar fact that a countable union of countable sets is countable. We formulate Zorn's Lemma (ZL) and show that it implies AC and the "Principle of Cardinal Comparability" (PCC, also known as the Law of Trichotomy). With the help of ZL, we establish the basic arithmetic of cardinalities of infinite sets. We give examples of the use of ZL in mathematics: maximal ideals in rings, bases for vector spaces. We give pointers to the literature for proofs that the statements involved (e.g. that every vector space has a basis and that any two bases of the same vector space have the same cardinality) actually *need* ZL for their proofs. We develop the theory of well-ordered sets with the principles of transfinite induction and recursion. In the last section of this chapter, we give rigorous proofs of the equivalences between AC, ZL, PCC and the Well-Ordering Theorem (which states that there is a well-ordering on every set).

In Chapter 2 ("Models"), we begin to do logic. Mathematical knowledge is organized in the form of *statements* (propositions, theorems, but also conjectures, questions), and logic aims to analyse at least two aspects of these statements: they can be true or false (in a particular mathematical interpretation), and they can sometimes be proved.

In order to study these aspects in a precise way, we restrict our attention to certain idealized, abstract statements which are just strings of symbols of a certain alphabet: the sentences of a formal language. We can then define what it means for such an abstract sentence to be "true" in a particular interpretation (for example, the sentence "2 is a square" is true in $\mathbb{R}$ but false in $\mathbb{N}$). Such an interpretation is called a *model*; this definition is by Alfred Tarski. The central theorem of this chapter is the *Compactness Theorem* which says that a set of axioms is compatible (i.e. "true" in at least *some* model) as soon as each of its finite subsets is. A proof of the Compactness Theorem using ultrafilters is provided; this proof can be skipped if the reader prefers to view the Compactness Theorem as a consequence of the Completeness Theorem, which is proved in Chapter 3. We study relations between models (isomorphisms, submodels and elementary submodels), and we embark on a little foray into *Model Theory*: the study of theories from the perspective of their classes of models. We prove quantifier elimination for algebraically closed fields, the Löwenheim–Skolem theorems and ω-categoricity of the theory of dense linear orders by Cantor's back-and-forth method.

Chapter 3 ("Proofs") defines abstract proofs: an abstract proof is a collection of sentences which is structured in such a way that every sentence which appears in it is either an assumption or can be seen as a direct consequence of sentences which have appeared "before" in the proof; we use the picture of a tree, and our proofs are so-called *natural deduction trees*. Every proof has a unique conclusion, which is a sentence. The theory of proofs takes up Chapter 3, *Proofs*. We prove the most fundamental theorem of logic, *Gödel's Completeness Theorem*, in this chapter. This theorem says that a sentence is always true (in all possible interpretations) precisely if it is the conclusion of such a proof tree. A small section deals with conservative extensions and Skolem theories.

Now that the student knows what a formal theory is (a collection of sentences in the sense of Chapters 2 and 3), we can look at the formal theory of sets. In Chapter 4 ("Sets Again"), we explain how the theory of sets can be set up with axioms; we obtain *Zermelo–Fraenkel Set Theory ZFC*. We define *ordinal numbers* and prove their basic properties. We can then give meaning to the notion of a *cardinal number*. We hope to convince you that this theory is sufficient for "doing mathematics", but actually we cannot (in the scope of this book) even scratch the surface of this vast topic.

These four chapters comprise just a first introduction to the vast and growing subject of mathematical logic. In the Appendix, we have tried to help the reader who wishes to know more by mentioning some important further developments in the 1900s, and by suggesting some books and expository articles for further reading. In particular, we have sketched the most important developments around Gödel's famous Incompleteness Theorems and in Computability Theory as it began with Alan Turing's foundational work.

Our treatment of proofs in Chapter 3 is very rudimentary, and only covers the introduction of a formal system so as to be able to state and prove the Completeness Theorem. But precisely because of the *Incompleteness Theorems*, Proof Theory is really about the comparison of such systems and about questions of relative consistency: can one "stronger" system prove that another is without contradiction? And, how does one measure the strength of a system? The study of proof systems from this point of view also has applications outside logic, in mathematics and in computer science.

Similar remarks apply to Model Theory; we have developed enough material in order to be able to prove the Completeness Theorem and have given some first results in Model Theory in the hope of giving the reader a flavour of the field. But Model Theory is a fast-developing area with many deep results in the theory itself as well as beautiful applications, for example to Algebraic Geometry. Some of these are described very briefly in the Appendix.

Our first chapter tries to make precise some aspects of what is sometimes called "naïve" set theory—the handling of sets as done by every mathematician on a daily basis. An axiomatic basis is presented in Chapter 4. Set theory as a field of mathematical logic is really the study of this system and related systems. What is the proof-theoretic strength of these systems, and how can one construct natural models: for example, different models where the Continuum Hypothesis fails, or where it holds? And, which of these systems are approximations of the "true" world of sets? Or is there no such true world? In the Appendix, the reader can find some pointers to the literature about these topics.

Contents

Chapter 1
Sets

This chapter intends to develop your understanding of sets beyond the use you have made of them in your first years of mathematics.

The first mathematician who thought about sets, and realized that it makes sense to organize mathematical knowledge using the concept of "set", was Georg Cantor (1845–1918). His name will appear at several places in this book. For biographical information on Cantor, whose genius did not receive proper recognition in his time and who had a troubled life, see [12] or the sketch in [37].

The first triumph of Cantor's theory of sets was that he could show that there are "different kinds of infinity": although the set of rational numbers and the set of irrational numbers are both infinite, there are "more" irrational numbers than rational ones.

An important part of this chapter explains how to calculate with these different kinds of infinity. It turns out that in order to set up the theory, it is necessary to adopt a principle which was first formulated by Ernst Zermelo (1871–1953) in 1904: the Axiom of Choice. In Sections 1.2 and 1.3, we introduce you to working with this axiom and with a useful equivalent principle: Zorn's Lemma.

Then, in Section 1.4, we develop another concept which originates with Cantor: that of a *well-order*. Thanks to this idea, we can extend proofs by induction along $\mathbb{N}$ to proofs for arbitrarily "large" sets.

Finally, in Section 1.5 precise proofs are given of the equivalences between various versions of the Axiom of Choice.

So let us start. Instead of trying to formulate what a "set" is, we assume that you already have some idea of it. A set has "elements". If X is a set and x is an element of X, we write $x \in X$. Think of X as a *property*, and the elements of X as the things having property X. For the negation of $x \in X$, we write $x \notin X$.

© Springer Nature Switzerland AG 2018
I. Moerdijk, J. van Oosten, *Sets, Models and Proofs*, Springer Undergraduate
Mathematics Series, https://doi.org/10.1007/978-3-319-92414-4_1

Georg Cantor

A set is completely determined by its elements. This means: suppose the sets X and Y have the same elements. So for all $x \in X$ we have $x \in Y$, and vice versa. Then we consider X and Y to be the same set: $X = Y$.

A set X is called a *subset* of a set Y if every element of X is also an element of Y. We write: $X \subseteq Y$. If we want to stress that $X \subseteq Y$ but $X \neq Y$ we may write $X \subsetneq Y$. To give an example of a subset: if $x \in X$ then there is a subset $\{x\}$ of X, which has only the one element x.

The notation $\{x\}$ is an example of the *brace notation* $\{\}$ for writing sets: we specify a set by giving its elements, either by listing them all (using dots if necessary, as in $\{0, 1, 2, \ldots\}$), or by giving the property that the elements of the set must satisfy, as in

$$\mathbb{R}_{>0} = \{x \mid x \in \mathbb{R} \text{ and } x > 0\}.$$

The following examples of sets are familiar to you: the *empty set* $\emptyset$, which has no elements, the set $\mathbb{N} = \{0, 1, \ldots\}$ of *natural numbers*, and likewise the sets $\mathbb{Z}$, $\mathbb{Q}$ and $\mathbb{R}$ of integers, rational numbers and real numbers, respectively. (By the way, it was Cantor who introduced the notation $\mathbb{R}$.) Let us make clear once again, to avoid misunderstanding:

in this book, the natural numbers start with 0.

Let us also recall the following basic operations on sets:

The *union* $X \cup Y$ of X and Y is the set $\{z \mid z \in X \text{ or } z \in Y\}$.

The *intersection* $X \cap Y$ is the set $\{z \mid z \in X \text{ and } z \in Y\}$.

The notations $\cup$ and $\cap$ are extended to (possibly infinite) indexed families of sets: $\bigcup_{i \in I} X_i$ and $\bigcap_{i \in I} X_i$. These are defined by:

$$\bigcup_{i \in I} X_i = \{x \mid \text{for some } i \in I, x \in X_i\};$$
$$\bigcap_{i \in I} X_i = \{x \mid \text{for all } i \in I, x \in X_i\}.$$

If $X \subseteq Y$, the *complement* of X in Y, written as $Y - X$, is the set of those elements of Y that are not elements of X (in the literature, one also finds the notation $Y \setminus X$).

The sets X and Y are *disjoint* if they have no elements in common. This is equivalent to: $X \cap Y = \emptyset$.

We also assume that you have an idea of what a *function* between sets is: a function f from X to Y (notation $f : X \to Y$) gives us for each element x of X a unique element $f(x)$ of Y, the value of the function f at x.

A function $f : X \to Y$ is completely determined by its values. That means: if f and g are functions from X to Y and for every $x \in X$ we have $f(x) = g(x)$, then f and g are the same function: $f = g$.

The following are examples of functions: for every set X, there is the *empty function* from $\emptyset$ to X, and the *identity function* from X to X (this function, say $I_X : X \to X$, satisfies $I_X(x) = x$ for every $x \in X$). Given functions $f : X \to Y$ and $g : Y \to Z$ there is the *composite* $g \circ f : X \to Z$ (or $gf : X \to Z$), which is defined by: $g \circ f(x) = g(f(x))$ for all $x \in X$.

In general, if X and Y are sets, and for every element x of X a subset Y_x of Y is given such that Y_x has exactly one element, then there is a (unique) function $f : X \to Y$ with the property that $f(x) \in Y_x$ for every $x \in X$.

Let us recall some more definitions.

A function $f : X \to Y$ is called *injective* (or 1–1) if for each $x, y \in X$ it holds that $f(x) = f(y)$ implies $x = y$.

The function f is *surjective* (or onto) if every $y \in Y$ is equal to $f(x)$ for some $x \in X$.

And f is called *bijective* if there is a function $g : Y \to X$ such that for all $x \in X$ and all $y \in Y$ the equalities $g(f(x)) = x$ and $f(g(y)) = y$ hold. Given f, the function g is unique if it exists, and is called the *inverse* of f, notation: f^{-1}.

If $f : X \to Y$ is a function, there is the subset of Y which consists of all elements of the form $f(x)$ for $x \in X$. This subset is called the *image* of the function f. Likewise, if A is a subset of Y, there is the subset of X which consists of all elements $x \in X$ such that $f(x) \in A$. This subset is sometimes denoted by $f^{-1}(A)$, and called the *inverse image* or *preimage* of A under f.

Exercise 1 Prove the following statements:

(a) A function $f : X \to Y$ is bijective if and only if it is both injective and surjective.
(b) A function $f : X \to Y$ is surjective if and only if the image of f is equal to Y.
(c) If $f : X \to Y$ is injective, then f is a bijective function from X to the image of f.

Exercise 2 Suppose that X is a set which can be written as a union $X = \bigcup_{i \in I} X_i$, and for every $i \in I$ a function $f_i : X_i \to Y$ is given in such a way that, whenever $X_i \cap X_j \neq \emptyset$, then $f_i(x) = f_j(x)$ for every $x \in X_i \cap X_j$. Prove that there is a unique function $f : X \to Y$ which is such that for all $i \in I$ and all $x \in X_i$, $f(x) = f_i(x)$.

1.1 Cardinal Numbers

A set X is *finite* if for some $n \in \mathbb{N}$ there is a bijective function $f : \{1, \ldots, n\} \to X$ (for $n = 0$, the set $\{1, \ldots, n\}$ is empty). This means that X has exactly n elements; we call n the *cardinality* of X and write $|X|$ for this number (in the literature, the notation $\sharp(X)$ is also sometimes used). A set which is not finite is called *infinite*.

Exercise 3 For an arbitrary set X there is at most one n such that $|X| = n$.

We introduce the following notations for (constructions on) sets:

- We assume that given sets X and Y, for every $x \in X$ and $y \in Y$ the *ordered pair* (x, y) is given, and that we have a set $X \times Y$, given as

$$X \times Y = \{(x, y) \mid x \in X, y \in Y\},$$

  which we call the *Cartesian product* of X and Y; there are projection functions $\pi_X : X \times Y \to X$ and $\pi_Y : X \times Y \to Y$ sending the pair (x, y) to x and to y, respectively; whenever we have functions $f : Z \to X$ and $g : Z \to Y$ there is a *unique* function $h : Z \to X \times Y$ (sending $z \in Z$ to the pair $(f(z), g(z))$) with the property that $\pi_X h = f$ and $\pi_Y h = g$;
- Given a set A and a natural number n, we have the set A^n of n-tuples of elements of A. If $n = 0$ this set will have exactly one element (the *empty tuple*); for $n > 0$, elements of A^n are denoted $(a_1, \ldots, a_n)$.

- The set $X + Y$ is the *disjoint sum* of X and Y, constructed as

$$\{(0, x) \mid x \in X\} \cup \{(1, y) \mid y \in Y\}.$$

- The set Y^X is the set of all functions $f : X \to Y$.
- The set $\mathcal{P}(X)$ is the *power set* of X, that is, the set of all subsets of X.

Exercise 4 Show the following equalities for finite sets X and Y:

(a) $|X \times Y| = |X| \times |Y|$.
(b) $|X + Y| = |X| + |Y|$.
(c) $|Y^X| = |Y|^{|X|}$.
(d) $|\mathcal{P}(X)| = 2^{|X|}$.

For arbitrary sets X and Y we write $X \sim Y$ to indicate that there is a bijective function from X to Y.

Exercise 5 Prove the following facts about $\sim$:

(a) $X \sim X$.
(b) If $X \sim Y$, then $Y \sim X$.
(c) If $X \sim Y$ and $Y \sim Z$, then $X \sim Z$.

We write $X \le Y$ if there is an injective function from X to Y.

Since every bijective function is injective, $X \sim Y$ implies $X \le Y$. Notice also that if $X' \sim X$ and $Y \sim Y'$, then $X' \le Y'$ whenever $X \le Y$.

In the literature the following theorem is sometimes called the "Schröder–Bernstein Theorem" and sometimes the "Cantor–Bernstein Theorem" (Felix Bernstein, 1878–1956; Ernst Schröder, 1841–1902).

Theorem 1.1.1 (Schröder–Cantor–Bernstein)   *If $X \le Y$ and $Y \le X$, then $X \sim Y$.*

Proof. Assume that $f : X \to Y$ and $g : Y \to X$ are injective functions; we construct a bijection from X to Y.

Define a subset A of X as follows: A consists of those $x \in X$ for which there exists an $x' \in X$ such that x' does not belong to the image of g, and

$$x = (gf)^n(x')$$

for some natural number n. The notation $(gf)^n(x')$ stands for: the function gf applied n times to x' (for $n = 0$, this is just x'). So, $x \in A$ holds precisely when *either* x is not in the image of g, *or* there is some $x'' \in A$ such that $x = gf(x'')$. For $x \notin A$, we must have $x = g(y)$ for a unique $y \in Y$ (since g is injective); denote this y by $g^{-1}(x)$.

Now define the function $h : X \to Y$ as follows:

$$h(x) = \begin{cases} g^{-1}(x) \text{ if } x \notin A \\ f(x) \ \text{ if } x \in A. \end{cases}$$

Let us first show that h is 1–1. Clearly, the restrictions of h to A and to $X - A$ are injective, so we must prove that when $x \in A$ and $x' \in X - A$, we cannot have $h(x) = h(x')$. But for $x' \in X - A$ we have that $gh(x') = g(g^{-1}(x')) = x'$, so $gh(x') \notin A$; whereas for $x \in A$ we have that $gh(x)$ is again an element of A. Therefore $gh(x) \neq gh(x')$, and thus $h(x) \neq h(x')$. We conclude that the function h is 1–1.

Now, we claim that h is onto. Pick $y \in Y$; we have to find an $x \in X$ with $h(x) = y$. Consider the element $g(y)$ of X. If $g(y) \notin A$, then $h(g(y)) = y$, so we can take $g(y)$ for x. If $g(y) \in A$, then (since clearly $g(y)$ belongs to the image of g) there is some $x \in A$ such that $g(y) = (gf)(x)$. Then by injectivity of g we have $y = f(x)$, so since $x \in A$, $y = h(x)$, and we are done. ∎

We extend the notation $|X|$ to arbitrary (not necessarily finite) sets X and use it as follows:

> We say $|X| = |Y|$ if $X \sim Y$.
> The notation $|X| \leq |Y|$ means the same as $X \leq Y$.
> We write $|X| < |Y|$ if $|X| \leq |Y|$ but not $|X| = |Y|$ (equivalently, by Theorem 1.1.1: $|X| \leq |Y|$ but not $|Y| \leq |X|$).

When $|X| < |Y|$ we think of X as "smaller than" Y; similarly, if $|X| \leq |Y|$ we think of X as "at most as large as" Y.

Definition 1.1.2 For a set X, we refer to $|X|$ as the *cardinality* of X. An object of the form $|X|$ is called a *cardinal number*.

We regard every $n \in \mathbb{N}$ as a cardinal number, namely $n = |\{1, \ldots, n\}|$. Note that this also means $0 = |\emptyset|$. Note also that $n \leq m$ as cardinal numbers if and only if $n \leq m$ in the usual ordering of $\mathbb{N}$. There are also infinite cardinal numbers, such as $|\mathbb{N}|$.

Definition 1.1.3 We have the following operations on cardinal numbers:

- $|X| \times |Y| = |X \times Y|$.
- $|X| + |Y| = |X + Y|$.
- $|Y|^{|X|} = |Y^X|$.
- $2^{|X|} = |\mathcal{P}(X)|$.

Exercise 6 Is this a correct definition? What do you have to check?

Exercise 7 Prove that the operations $+$, $\times$ and $(-)^{(-)}$ for cardinal numbers satisfy the following usual rules of arithmetic:

(a) $(|X| + |Y|) \times |Z| = (|X| \times |Z|) + (|Y| \times |Z|)$.
(b) $|X|^{|Y|+|Z|} = (|X|^{|Y|}) \times (|X|^{|Z|})$.
(c) $(|X| \times |Y|)^{|Z|} = (|X|^{|Z|}) \times (|Y|^{|Z|})$.
(d) $(|X|^{|Y|})^{|Z|} = |X|^{|Y| \times |Z|}$.

Formulate and prove some more of these rules yourself.

Now let us consider the cardinalities of power sets.

There is a bijective function from $\mathcal{P}(X)$ to $\{0, 1\}^X$: for a subset $S \subseteq X$ we define the *characteristic function* of S by

$$\chi_S(x) = \begin{cases} 1 \text{ if } x \in S \\ 0 \text{ if } x \notin S. \end{cases}$$

Conversely, every function $\chi : X \to \{0, 1\}$ is of the form χ_S for a unique subset S of X, namely $S = \{x \in X \mid \chi(x) = 1\}$.
Therefore, $|\mathcal{P}(X)| = |\{0, 1\}^X| = 2^{|X|}$.

Proposition 1.1.4 *The following equalities between cardinal numbers hold:*

  (i) $|\mathbb{N}| = |\mathbb{N}| + |\mathbb{N}|$.
 (ii) $|\mathbb{N}| = |\mathbb{N}| \times |\mathbb{N}|$.
(iii) $|\mathbb{N}| = |\mathbb{Z}| = |\mathbb{Q}|$.

Proof. We indicate only a proof of (ii), leaving the other statements as exercises. A bijection $f : \mathbb{N} \to \mathbb{N} \times \mathbb{N}$ is indicated in the following diagram:

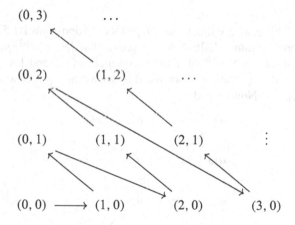

The path obtained by following the arrows indicates the successive values of f. Thus $f(0) = (0, 0)$, $f(1) = (1, 0)$, etcetera. ∎

Exercise 8 Find a formula for the inverse of the function f indicated in the proof above. Also give proofs of the other statements of Proposition 1.1.4.

Proposition 1.1.5 (Cantor) *For every set A we have the strict inequality*

$$|A| < 2^{|A|}.$$

In other words, there is an injective function $A \to \mathcal{P}(A)$ but there is no bijective function between these sets.

Notice that you already know Proposition 1.1.5 for finite sets; indeed, $n < 2^n$ is true for every natural number n.

Proof. It is easy to construct the required injective function $f : A \to \mathcal{P}(A)$. Just define $f(a) = \{a\}$ (the *singleton set* whose only element is a).

For the statement in the proposition, we shall show something stronger than required, namely that there cannot be any surjective function from A to $\mathcal{P}(A)$. The argument we use is known as the *Cantor diagonal argument*. Suppose that

$$s : A \to \mathcal{P}(A)$$

is surjective. We can construct a subset D of A by putting

$$D = \{a \in A \mid a \notin s(a)\}.$$

Since s is assumed surjective, there must be some $a_0 \in A$ with $s(a_0) = D$. But now the simple question 'does a_0 belong to D?' gets us into trouble: we have

$$a_0 \in D \text{ iff } a_0 \notin s(a_0) \text{ (by definition of } D\text{)}$$
$$\text{iff } a_0 \notin D \quad \text{(since } D = s(a_0)\text{)}.$$

Thus, our assumption that such a surjective function s exists leads to a contradiction. ∎

Example 1.1.6 This example illustrates the proof of Proposition 1.1.5 and explains the term 'diagonal argument'. In order to prove that $|\mathbb{N}| < 2^{|\mathbb{N}|}$, suppose for a contradiction that the set $\{0, 1\}^{\mathbb{N}}$ of infinite sequences of 0-s and 1-s is in bijective correspondence with $\mathbb{N}$. Then we can list this set as $a_0, a_1, \ldots$, where a_i is the sequence $a_{i0}, a_{i1}, \ldots$. Now consider:

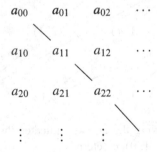

$$a_{00} \quad a_{01} \quad a_{02} \quad \cdots$$

$$a_{10} \quad a_{11} \quad a_{12} \quad \cdots$$

$$a_{20} \quad a_{21} \quad a_{22} \quad \cdots$$

$$\vdots \qquad \vdots \qquad \vdots$$

Clearly, the sequence

$$(1 - a_{00}), (1 - a_{11}), (1 - a_{22}), \ldots$$

does not appear in the list, contradicting the assumption that we were listing *all* 01-sequences. You should convince yourself that this pictorial argument is essentially the same as the more general one of the proof of Proposition 1.1.5.

Proposition 1.1.5 has an important consequence: there are infinitely many infinite cardinal numbers. In fact, if we write $|\mathbb{N}| = \omega$ as is customary, we have

$$\omega < 2^{\omega} < 2^{(2^{\omega})} < \cdots$$

Let us try to determine the position of some familiar sets from analysis from the point of view of their cardinal numbers. We have already seen that the cardinal numbers of $\mathbb{N}$, $\mathbb{Q}$ and $\mathbb{Z}$ are the same (Proposition 1.1.4). These are so-called *countable sets*. We make the following definition:

Definition 1.1.7 A set X is called *countable* if X is empty or there is a surjective function $\mathbb{N} \to X$. We say that X is *countably infinite* if it is countable and infinite.

So, if a non-empty set X is countable, one can 'enumerate' all its elements as

$$X = \{x_0, x_1, x_2, \ldots\}$$

(but repetitions may occur).

Exercise 9

(a) Show that if $f : \mathbb{N} \to X$ is surjective, there is a function $g : X \to \mathbb{N}$ such that $f(g(x)) = x$ for all $x \in X$. How do you define $g(x)$? Conclude that $|X| \leq |\mathbb{N}|$.
(b) Show that every subset of a countable set is countable. Conclude from this and item (a) that a set X is countable precisely when $X \leq \mathbb{N}$.
(c) Show that if X is countable then X is finite or $|X| = \omega$.
(d) Show that if X and Y are countable, so are $X \times Y$ and $X + Y$.

An example of an *uncountable* set is $\{0, 1\}^{\mathbb{N}}$, as follows from Proposition 1.1.5.

What about the real numbers? There are several ways to determine the cardinality of $\mathbb{R}$. Our approach uses the so-called *Cantor set*, a subset C of $\mathbb{R}$ that was defined by Cantor in order to prove that $\mathbb{R}$ is not countable. However, the set C has a lot of independent interest and is also often used in topology. It is constructed as the intersection

$$C = \bigcap_{n \in \mathbb{N}} C_n$$

of an infinite sequence of smaller and smaller subsets of $\mathbb{R}$,

$$\mathbb{R} \supsetneq C_0 \supsetneq C_1 \supsetneq C_2 \supsetneq \cdots$$

Each C_i is a union of closed intervals. C_0 is the interval $[0, 1]$, and C_{n+1} is obtained from C_n by "cutting out the middle third" from each of the intervals which make up C_n:

$$C_1 = [0, \tfrac{1}{3}] \cup [\tfrac{2}{3}, 1]$$
$$C_2 = [0, \tfrac{1}{9}] \cup [\tfrac{2}{9}, \tfrac{1}{3}] \cup [\tfrac{2}{3}, \tfrac{7}{9}] \cup [\tfrac{8}{9}, 1]$$
etc.

Thus a point p of C is uniquely determined by specifying for each n the interval of C_n to which p belongs. We can code this specification as a sequence of 0's and

1's where 0 means "left" and 1 means "right" (for each subinterval of C_n, there are exactly two subintervals of C_{n+1} contained in it). For example,

$$p = 010 \cdots$$

is the point which lies in the left part $[0, \frac{1}{3}]$ of C_1, then in the right part $[\frac{2}{9}, \frac{1}{3}]$ of the two intervals of C_2 contained in $[0, \frac{1}{3}]$, then in the left part $[\frac{6}{27}, \frac{7}{27}]$ at the next stage, etcetera. Since the length of the intervals tends to zero, the sequence p defines a unique element of C. In this way, we obtain a bijective function

$$\varphi : \{0, 1\}^{\mathbb{N}} \to C.$$

Thus,

$$|C| = 2^{\omega}.$$

Although C is just a subset of $\mathbb{R}$, the two sets are equally large in some sense:

Proposition 1.1.8 $|C| = |\mathbb{R}|$.

Proof. By Theorem 1.1.1, it suffices to prove that $|C| \leq |\mathbb{R}|$ and $|\mathbb{R}| \leq |C|$. Since C is a subset of $\mathbb{R}$ we obviously have

$$|C| \leq |\mathbb{R}|.$$

There are many ways to prove the converse inequality. For example, each real number x is determined by the set of rational numbers which are strictly below x. This defines an injective function

$$\psi : \mathbb{R} \to \mathcal{P}(\mathbb{Q}).$$

Since $|\mathbb{Q}| = |\mathbb{N}|$, hence $|\mathcal{P}(\mathbb{Q})| = 2^{\omega}$, we see that

$$|\mathbb{R}| \leq 2^{\omega}.$$

Since $|C| = 2^{\omega}$, we are done.   ∎

Exercise 10 Show that $2^{\omega} \times 2^{\omega} = 2^{\omega}$. Conclude that the field of complex numbers has the same cardinality as $\mathbb{R}$.

Exercise 11 Prove that for a subset A of $\mathbb{R}$, if $|A| = \omega$ then $|\mathbb{R} - A| = 2^{\omega}$. [Hint: use that $\mathbb{R} \sim \{0, 1\}^{\mathbb{N}}$. For countable $A \subseteq \{0, 1\}^{\mathbb{N}}$, show that $\{0, 1\}^{\mathbb{N}} \leq \{0, 1\}^{\mathbb{N}} - A$. Alternatively, given a countable subset A of $\mathbb{R}$, construct a Cantor-like set which does not meet A.]

Conclude that $C \sim P$, where P is the set of irrational numbers.

Exercise 12 Prove that $|\mathbb{R}^{\mathbb{N}}| = |\mathbb{R}|$.

Exercise 13 Let Cont denote the set of continuous functions $\mathbb{R} \to \mathbb{R}$.

(a) Show that the function Cont $\to \mathbb{R}^{\mathbb{Q}}$, which sends a continuous function $f : \mathbb{R} \to \mathbb{R}$ to its restriction to $\mathbb{Q}$, is injective.

(b) Prove that $|\text{Cont}| = |\mathbb{R}|$.

Exercise 14 Let $\mathcal{A} \subseteq \mathbb{N}^{\mathbb{N}}$ be a set of functions with the following property: for every $f \in \mathbb{N}^{\mathbb{N}}$ there is a function $g \in \mathcal{A}$ which satisfies: for all $n \in \mathbb{N}$ there exists an $m \geq n$ with $f(m) < g(m)$.

Prove that the set $\mathcal{A}$ is uncountable.

Exercise 15 Prove that the set of bijections $\mathbb{N} \to \mathbb{N}$ is uncountable.

Exercise 16 Call a function $f : \mathbb{N} \to \mathbb{N}$ *repetitive* if for every finite sequence of natural numbers $(a_1, \ldots, a_n)$ there exist a number $k \in \mathbb{N}$ satisfying

$$f(k) = a_1, \ f(k+1) = a_2, \ \ldots, \ f(k+n-1) = a_n.$$

(a) Show that if f is repetitive then for any $(a_1, \ldots, a_n)$ there are actually *infinitely many* numbers k with this property.

(b) Show that there are uncountably many repetitive functions $\mathbb{N} \to \mathbb{N}$. [Hint: it may help to use Exercise 15.]

Exercise 17 For each of the following sets of functions, determine whether it is finite, countably infinite or uncountable. Give an argument in each case.

(a) $A = \{f : \mathbb{N} \to \mathbb{N} \,|\, \text{for all } n, \ f(n) = f(n+1) + f(n+2)\}$.

(b) $D = \{f : \mathbb{N} \to \mathbb{N} \,|\, \text{for all } n, \ f(n)^2 - 3f(n) + 2 = 0\}$.

(c) $C = \{f : \mathbb{N} \to \mathbb{N} \,|\, \text{for all } n, \ f(n) \geq f(n+1)\}$.

1.1.1 The Continuum Hypothesis

Suppose A is a subset of $\mathbb{R}$ such that $\mathbb{N} \subseteq A \subseteq \mathbb{R}$. Then we know that

$$\omega \leq |A| \leq 2^{\omega} = |\mathbb{R}|$$

and at least one of the inequalities must be strict because $\omega < 2^{\omega}$. We may ask ourselves: can it happen that *both* inequalities are strict? Is there a subset A of $\mathbb{R}$ such that

$$\omega < |A| < 2^{\omega}?$$

This problem was raised by Cantor. Unable to find such a set, he formulated the so-called *Continuum Hypothesis*, which states that every subset of $\mathbb{R}$ which contains $\mathbb{N}$ is either countable or has cardinality 2^{ω}.

It cannot be decided on the basis of the axioms of Set Theory (see Chapter 4) whether the Continuum Hypothesis (CH) is true or false. Kurt Gödel (1906–1978) showed in 1940 ([24]) that CH does not contradict these axioms; on the other hand, Paul Cohen (1934–2007) proved in 1963 ([9]) that its negation doesn't either. This means: one cannot derive a contradiction by logical reasoning on the basis of CH and the axioms of Set Theory, but it is also impossible to prove CH from these axioms.

Kurt Gödel was already famous for his "Incompleteness Theorems" when he proved the "Consistency of the Continuum Hypothesis". Paul Cohen's result, usually referred to as "Independence of the Continuum Hypothesis", solved a problem posed by Hilbert in 1900 (see also the Introduction), and won him the Fields Medal in 1966. The Fields Medal is, in Mathematics, what the Nobel Prize is for Physics and other sciences. Cohen's is the only Fields Medal ever awarded for work in Logic.

1.2   The Axiom of Choice

An axiom in mathematics is a statement or a principle of reasoning that is simply assumed because it is intuitively correct but it cannot be proved from more basic facts. An example of such an axiom is the principle of *mathematical induction* for natural numbers (see also the statements just before Proposition 1.4.4).

Of course, in general it is far from easy to see that a principle 'cannot be proved': for over 2000 years, mathematicians tried to prove that Euclid's controversial "Parallel postulate" could be proved from the other axioms in geometry, until it was established beyond doubt in the nineteenth century that this axiom does not follow from the other four axioms of Euclid.

The Axiom of Choice is a bit peculiar among the axioms of mathematics, because it asserts the *existence* of a function, without telling you what it is. It takes a while to get used to the axiom, and it has remained somewhat controversial ever since its formulation by Zermelo in 1904. Nevertheless, modern mathematics is unthinkable without it, and almost all mathematicians accept it as true. Moreover, another famous result of Gödel (also in [24]) asserts that the Axiom of Choice does not contradict the other axioms of Set Theory (we will see these axioms in Chapter 4). Eventually, it was again Paul Cohen who showed that the Axiom of Choice does not follow from the other axioms of set theory either ([10]).

Informally, the Axiom of Choice states that given a collection of non-empty sets, there is a way to choose an element from each set in the collection. Here is a more precise formulation, which looks simpler.

Definition 1.2.1 The *Axiom of Choice* (AC) is the assertion that for every surjective function $f : X \to Y$ there exists a "section"; that is, a function $s : Y \to X$ such that $f(s(y)) = y$ for each $y \in Y$.

In order to "define" such a section as in Definition 1.2.1, one has to "choose", for each $y \in Y$, an $x \in X$ such that $f(x) = y$. In general, the Axiom of Choice is needed when:

- there is more than one x such that $f(x) = y$ (see Exercise 20), and
- Y is infinite (see Exercise 21).

But even in these circumstances the Axiom of Choice is not *always* necessary to obtain a section. For example, if, in Definition 1.2.1, $X = \mathbb{N}$, we can simply define $s(y)$ as the *least n* such that $f(n) = y$ (this is the solution of Exercise 9 (i)).

An example of a genuine application of the Axiom of Choice is given by the following simple proposition, which you may have thought was self-evident.

Proposition 1.2.2 *If X is an infinite set, then there is an injective function $\mathbb{N} \to X$, hence $\omega \leq |X|$.*

Proof. Intuitively, one can "choose" for each $n \in \mathbb{N}$ an element $g(n) \in X$ such that $g(n)$ is different from all elements chosen before. This reasoning is essentially correct, but below we present a detailed proof, just in order to make clear exactly how the Axiom of Choice is used.

First we remark that if $(x_1, \ldots, x_n)$ is a finite sequence of elements of X such that $x_i \neq x_j$ whenever $i \neq j$, then there is an element $x_{n+1} \in X$ such that $x_i \neq x_{n+1}$ for all $i \leq n$; for if not, we would have $|X| = n$, and X would be finite.

Now let A be the set of all such finite sequences $(x_1, \ldots, x_n)$ with at least one element; and let B be the union $A \cup \{*\}$, where $*$ is any element not in A. Define a function $f : A \to B$ by:

$$f((x_1)) = *$$
$$f((x_1, \ldots, x'_{n+1})) = (x_1, \ldots, x_n).$$

Then by our remark, we see that $f : A \to B$ is surjective, and so the Axiom of Choice says there is a section $s : B \to A$.

This section s allows us to define a function $g : \mathbb{N} \to X$ by induction: let $g(0)$ be the element of X such that $s(*) = (g(0))$; if $g(0), \ldots, g(n)$ have been defined, let $g(n + 1)$ be the element of X such that

$$s((g(0), \ldots, g(n))) = (g(0), \ldots, g(n + 1)).$$

Convince yourself that the function g thus defined is indeed injective. ∎

Remark 1.2.3 That Proposition 1.2.2 is a *genuine* application of AC (that is, it cannot be proved without AC) follows from Theorem 10.1 in [38].

Exercise 18 Let SCB^{op} (the "opposite Schröder–Cantor–Bernstein theorem") be the statement that given *surjective* functions $X \to Y$ and $Y \to X$, we have $X \sim Y$. In this exercise we show that SCB^{op} follows from AC but cannot be proved without it. The argument relies on Remark 1.2.3 and is due to Joel Hamkins.

(a) Use AC to prove SCB^{op}.

(b) Let X be an infinite set. Let A be the set of all (not just nonempty) finite strings of elements of X in which each element appears only once. Let $B = A \cup \{*\}$, where $* \notin A$. Show that there are surjective functions $A \to B$ and $B \to A$.

(c) Let A and B be as in (b). Suppose $A \sim B$. Show that there is a 1–1 function $\mathbb{N} \to A$.

(d) For A, X as in (b) and a 1–1 function $\mathbb{N} \to A$, show that there is a 1–1 function $\mathbb{N} \to X$.

(e) Conclude that SCB^{op} implies the statement of Proposition 1.2.2.

Exercise 19 Use Proposition 1.2.2 to show that if A is infinite and B is countable,

$$|A| + |B| = |A|.$$

Exercise 20 If A is nonempty and $s : A \to B$ is injective, there is a surjective function $f : B \to A$ such that $f(s(a)) = a$ for all $a \in A$. Prove this without using the Axiom of Choice.

Exercise 21 Prove the Axiom of Choice (every surjective $f : X \to Y$ has a section) in the following two special cases:

(a) Y is finite. [Hint: induction on the cardinality of Y.]

(b) X is countable.

The Axiom of Choice is essential for deriving basic properties of cardinalities, such as given by the following proposition.

Proposition 1.2.4 *Let I be a countable set and suppose for each $i \in I$, a countable set X_i is given. Then the union*

$$\bigcup_{i \in I} X_i$$

is again a countable set.

Proof. If $I' \subseteq I$ is the subset $\{i \in I \mid X_i \neq \emptyset\}$ then $\bigcup_{i \in I} X_i = \bigcup_{i \in I'} X_i$, and I' is countable by Exercise 9 (iv). So we may as well assume that X_i is nonempty for each $i \in I$.

If I is empty, the union is empty, hence countable. So let I be nonempty.

Let $g : \mathbb{N} \to I$ be a surjective function; such g exists because I is countable.

Let J be the set of all pairs (f, i) such that $f : \mathbb{N} \to X_i$ is surjective. The function $J \to I$, given by $(f, i) \mapsto i$, is surjective, because each X_i is nonempty and countable. By AC, it has a section s. Let $f_i : \mathbb{N} \to X_i$ be such that $s(i) = (f_i, i)$.

Now consider the function

$$h : \mathbb{N} \times \mathbb{N} \to \bigcup_{i \in I} X_i$$

defined by: $h(n, m) = f_{g(n)}(m)$. Convince yourself that h is surjective. Composing h with a surjective function $\mathbb{N} \to \mathbb{N} \times \mathbb{N}$, we see that $\bigcup_{i \in I} X_i$ is indeed countable, as required.   ∎

Remark 1.2.5 No doubt you have seen Theorem 1.2.4 before, but it may not be clear to you why the Axiom of Choice is *necessary* for its proof. The reason is, that in order to do the construction in the proof we have to *choose* surjective functions $\mathbb{N} \to X_i$ for all (possibly infinitely many) i. Indeed, without the Axiom of Choice we cannot prove that $\mathbb{R}$ (which is always an uncountable set, as we have seen) is not a union of countably many countable sets ([74])!

Exercise 22 Prove that the set

$$\{x \in \mathbb{R} \mid \sin(x) \in \mathbb{Q}\}$$

is countable.

Another simple application of the Axiom of Choice is in the following familiar theorem of analysis: if A is a bounded, infinite subset of $\mathbb{R}$, then there is an element $a \in A$ such that $A - \{a\}$ contains a sequence which converges to a (Bolzano–Weierstraß).

Later we shall see that, as a consequence of AC, we have for any two sets X and Y: either $|X| \leq |Y|$ or $|Y| \leq |X|$.

There are many statements which are equivalent to the Axiom of Choice. We shall now present one which is closer to our intuitive description of AC at the beginning of this section. We need the following definitions:

Definition 1.2.6 Let I be a set and let X_i be a set for each $i \in I$.

(a) The *disjoint sum* $\coprod_{i \in I} X_i$ is the set of all pairs (i, x) with $i \in I$ and $x \in X_i$.
(b) The *product* $\prod_{i \in I} X_i$ is the set of functions $f : I \to \bigcup_{i \in I} X_i$ which satisfy $f(i) \in X_i$ for each $i \in I$.

Proposition 1.2.7 *The Axiom of Choice is equivalent to the statement:*

(Π)   *For every family of sets $\{X_i \mid i \in I\}$ such that X_i is nonempty for each $i \in I$, the set*

$$\prod_{i \in I} X_i$$

is nonempty.

Proof. First we show that AC implies the statement (Π). So let X_i be nonempty for each i. Then the function $\coprod_{i \in I} X_i \to I$, which takes (i, x) to i, is surjective and therefore has a section s by AC.

Let

$$t : I \to \bigcup_{i \in I} X_i$$

be such that $s(i) = (i, t(i))$; then t is an element of $\prod_{i \in I} X_i$, as is easy to check.

Conversely, assume (Π) and let $f : X \to Y$ be a surjective function. Then for each $y \in Y$ the set $X_y = \{x \in X \mid f(x) = y\}$ is nonempty. By (Π), the set $\prod_{y \in Y} X_y$ is nonempty. But any element of this set is a section of f. ∎

Example 1.2.8 This example is meant to give some intuition about the use or non-use of AC. Consider the sets $\mathbb{R}$, $\mathbb{Z}$ and $\mathbb{Q}$. We have the equivalence relations $\sim_{\mathbb{Z}}$ and $\sim_{\mathbb{Q}}$ on $\mathbb{R}$:

$$x \sim_{\mathbb{Z}} y \text{ iff } y - x \in \mathbb{Z}$$
$$x \sim_{\mathbb{Q}} y \text{ iff } y - x \in \mathbb{Q}$$

and write $\mathbb{R}/\mathbb{Z}$ and $\mathbb{R}/\mathbb{Q}$, respectively, for the sets of equivalence classes. There are evident surjective functions

$$\varphi : \mathbb{R} \to \mathbb{R}/\mathbb{Z} \quad \text{and} \quad \psi : \mathbb{R} \to \mathbb{R}/\mathbb{Q}.$$

For φ, we can explicitly describe a section $\sigma : \mathbb{R}/\mathbb{Z} \to \mathbb{R}$: for every equivalence class ξ, the intersection of ξ with the half-open interval $[0, 1)$ contains exactly one point, which we take as $\sigma(\xi)$.

We cannot do something similar for ψ. The Axiom of Choice says that there must be a section, but it cannot be described explicitly. This is because the image of such a section is a subset of $\mathbb{R}$ which is not measurable, and that such sets cannot be explicitly defined is a result of [73].

The following puzzle has several versions; the one presented here is taken from [30].

Exercise 23 A cruel prison warden lines up countably many prisoners: $p_0, p_1, p_2, \ldots$ Every prisoner has received a cap on his head, which is either black or white. Every prisoner can see all prisoners with a higher number than his own. Eventually, each prisoner must guess the colour of his own cap (and of course, his life depends on it). Show that by cunningly using the Axiom of Choice, the prisoners can devise a strategy which will ensure that *only finitely many prisoners will make a wrong guess*.

1.3 Partially Ordered Sets and Zorn's Lemma

Zorn's Lemma (formulated independently by Kazimierz Kuratowski (1896–1980) and Max Zorn (1906–1993)) is a principle which is equivalent to the Axiom of Choice (as we shall see in Section 1.5), but formulated quite differently; in many cases, it is easier to apply than AC.

The formulation uses the notion of a *chain in a partially ordered set*, which we shall define first.

Definition 1.3.1 A *partially ordered set* or *poset* is a set P together with a relation $\leq$ between elements of P such that the following conditions are satisfied:

(i) For every $p \in P$, $p \leq p$ holds (one says that the relation $\leq$ is "reflexive").
(ii) For every $p, q, r \in P$, if $p \leq q$ and $q \leq r$ hold, then $p \leq r$ holds (the relation $\leq$ is said to be "transitive").
(iii) For every $p, q \in P$, if both $p \leq q$ and $q \leq p$ hold then $p = q$ (the relation $\leq$ is "antisymmetric").

We shall usually denote a poset as $(P, \leq)$, and the relation $\leq$ is pronounced as "less than or equal to". The converse relation $\geq$, "greater than or equal to", is defined by $x \geq y$ if and only if $y \leq x$.

Examples.

(a) The most important example of a poset is the powerset $\mathcal{P}(A)$ of a set A: the relation $p \leq q$ holds for subsets p and q of A if and only if p is a subset of q (i.e., $p \subseteq q$).
(b) If $(P, \leq)$ is a poset and $S \subseteq P$ then clearly the restriction of the relation $\leq$ to elements of S gives a poset $(S, \leq)$.
(c) Combining (a) and (b), we see that any collection $\mathcal{C}$ of subsets of a set X (i.e., $\mathcal{C} \subseteq \mathcal{P}(X)$) is naturally a poset when ordered by inclusion. Every poset $(P, \leq)$ is isomorphic to a poset of this form: consider the mapping from P to $\mathcal{P}(P)$ given by $x \mapsto \{y \in P \mid y \leq x\}$.
(d) The usual order relations on $\mathbb{N}$, $\mathbb{Z}$, $\mathbb{Q}$ and $\mathbb{R}$ make these sets into posets. These posets have the additional property that every two elements are comparable; that is, for each x and y, we have either $x \leq y$ or $y \leq x$. Posets in which every two elements are comparable are called *total* or *linear* orders. Note, that the poset $\mathcal{P}(X)$ is not a total order (at least if X has more than one element).
(e) Note, that if $(P, \leq)$ is a poset, then $(P, \geq)$ is a poset too.

Definition 1.3.2 Let $(P, \leq)$ be a poset.

(i) A subset C of P is called a *chain* if C with the restricted order is a total order. In other words, if either $p \leq q$ or $q \leq p$ holds, for any two elements p, q of C.
(ii) If S is any subset of P, an element p of P is called an *upper bound* for S if for each $s \in S$ we have $s \leq p$ (p itself doesn't need to be a member of S).
(iii) An element $p \in P$ is called *maximal* if no element is strictly greater: whenever $p \leq q$ we must have $p = q$.

Example 1.3.3 Let A be a fixed set with more than one element, and let P be the poset $\{S \subseteq A \mid S \neq A\}$, ordered by inclusion. This poset P has many maximal elements, namely the sets $A - \{a\}$ for $a \in A$. On the other hand, P does not have a greatest element. If $C \subseteq P$ is a chain and the union

$$\bigcup C = \{x \in A \mid \exists S \in C \ x \in S\}$$

is not equal to A, then this set is an upper bound for C. If $\bigcup C = A$, the chain C does not have an upper bound in P.

Exercise 24 Suppose X and Y are sets. Let P be the set of all pairs (A, f) where A is a subset of X and f is a function $A \to Y$. Then P is a poset with the following relation: $(A, f) \leq (B, g)$ iff $A \subseteq B$ and f is the restriction of g to A.
Show that if $C = \{(A_i, f_i) \mid i \in I\}$ is a chain in P, there is a unique function $f :$ $\bigcup_{i \in I} A_i \to Y$ such that for each i the function f_i is the restriction of f to A_i. Conclude that every chain in P has an upper bound in P.

Definition 1.3.4 *Zorn's Lemma* is the following assertion: if $(P, \leq)$ is a poset with the property that every chain in P has an upper bound in P, then P has a maximal element.

Note that if P satisfies the hypothesis of Zorn's Lemma, then P is nonempty. This is so because the empty subset of P is always a chain. However, checking that every chain has an upper bound in P usually involves checking this for the empty chain separately; that is, checking that P is nonempty (see Example 1.3.5 below).

Zorn's Lemma isn't a lemma, but a "principle" of a status similar to that of the Axiom of Choice (cf. the remarks at the beginning of Section 1.2).

Example 1.3.5 Let us look at maximal ideals in rings. Let R be a commutative ring with 1. As always in this book, for a ring we assume that $1 \neq 0$. Let P be the poset of all proper ideals of R (that is, ideals $I \neq R$), ordered by inclusion. If C is a nonempty chain of ideals, its union $\bigcup C$ is an ideal too, and $\bigcup C$ is proper, since $1 \notin \bigcup C$ because C consists of proper ideals. Moreover, P is nonempty since $\{0\}$ is a proper ideal by our assumption. So every chain in P has an upper bound. Hence, by Zorn's Lemma, P has a maximal element, which is a maximal ideal in R.

The existence of maximal ideals in rings cannot be proved *without* using Zorn's Lemma (or the Axiom of Choice). Other important basic facts from ring theory likewise need some form of the Axiom of Choice, such as the statement that every principal ideal domain has a maximal ideal, and the statement that every principal ideal domain is a unique factorization domain (see [36]).

Example 1.3.6 A similar example is that of bases for vector spaces. Let V be a vector space over $\mathbb{R}$ (or, in fact, any other field), for example the set of continuous functions from $[0, 1]$ into $\mathbb{R}$. Then V has a *basis*, that is, a subset $B \subseteq V$ with the property that every $v \in V$ can be written as a finite sum

$$v = c_1 b_1 + \cdots + c_n b_n$$

with $c_1, \ldots, c_n \in \mathbb{R}$ and $b_1, \ldots, b_n \in B$, and moreover this finite sum is *unique*. In order to prove this, let P be the poset of those subsets $B \subset V$ which are linearly independent (no element of B can be written as a linear combination of other elements of B), ordered by inclusion. If B is a maximal element of P, B must be a basis (check!).

Our next example uses Zorn's Lemma to prove the Axiom of Choice.

Proposition 1.3.7 *Zorn's Lemma implies the Axiom of Choice.*

As we have already remarked, Zorn's Lemma is equivalent to AC. Here we prove the most important implication. For the other direction, see Section 1.5.

Proof. We assume that Zorn's Lemma is true.

Suppose given a surjective function $f : X \to Y$. A *partial section* of f is a pair (A, u) where A is a subset of Y and $u : A \to X$ a function such that $f(u(y)) = y$ for each $y \in A$. Given two such partial sections (A, u) and (B, v), put $(A, u) \leq (B, v)$ iff $A \subseteq B$ and u is the restriction of v to A. Let P be the set of partial sections (A, u) of f; then with the relation $\leq$, P is a poset, as is easy to see.

P is nonempty; this is left to you. Moreover, if $C = \{(A_i, u_i) \mid i \in I\}$ is a chain in P, C has an upper bound; this is similar to Exercise 24. So the poset $(P, \leq)$ satisfies the hypotheses of Zorn's Lemma, and thus has a maximal element (A, s).

We claim that $A = Y$, and therefore that s is a section for f. Suppose that $y \notin A$. Then since f is surjective, there is an element $x \in X$ such that $f(x) = y$. If we define the function $s' : A \cup \{y\} \to X$ by

$$s'(w) = \begin{cases} s(w) & \text{if } w \in A \\ x & \text{if } w = y \end{cases}$$

then we see that $(A \cup \{y\}, s')$ is a partial section of f which is strictly greater than (A, s); this contradicts the fact that (A, s) is *maximal*. It follows that $A = Y$ and we have found a section for f. ∎

It is another important consequence of Zorn's Lemma that any two cardinal numbers $|X|$ and $|Y|$ can be compared: given sets X and Y, we have either $|X| \leq |Y|$ or $|Y| \leq |X|$. In other words, there is an injective function $X \to Y$, or there is an injective function $Y \to X$ (or both, of course).

Proposition 1.3.8 *Zorn's Lemma implies the following statement: for any two sets X and Y,*

$$|X| \leq |Y| \quad or \quad |Y| \leq |X|$$

holds.

The statement in the proposition is sometimes called the "Law of Trichotomy" (because one can equivalently put it as: one of the three possibilities $|X| < |Y|$,

$|X| = |Y|$ or $|Y| < |X|$ is true). We shall refer to it as the *Principle of Cardinal Comparability* (PCC).

Conversely, the Principle of Cardinal Comparability can be shown to be equivalent to Zorn's Lemma (or AC). For this, see Section 1.5.

Proof. Let X and Y be sets. We consider a poset P of triples (U, f, V), where $U \subseteq X$, $V \subseteq Y$ and $f : U \to V$ is a bijective function. These triples are ordered in a way reminiscent of the proof of Proposition 1.3.7: $(U, f, V) \leq (U', f', V')$ iff $U \subseteq U'$ and f is the restriction of f' to U (note, that this implies that $V \subseteq V'$).

P is nonempty, since we have $\emptyset$ as subset of both X and Y, and the "empty function" is bijective.

If $\{(U_i, f_i, V_i) \mid i \in I\}$ is a chain in P, there is a well-defined function $f : \bigcup_{i \in I} U_i \to \bigcup_{i \in I} V_i$ which is a bijection. Therefore, every chain in P has an upper bound, and by Zorn's Lemma P has a maximal element (U, f, V). If $U \neq X$ and $V \neq Y$, say $x \in X - U$ and $y \in Y - V$, we can obviously define a bijection between $U \cup \{x\}$ and $V \cup \{y\}$ which extends f, and this contradicts the maximality of (U, f, V). Hence *either* $U = X$, in which case f is an injective function from X into Y, *or* $V = Y$, in which case the inverse of f is an injective function $Y \to X$.∎

Exercise 25 Prove the following variation: if X and Y are nonempty sets, then there is either a surjective function $X \to Y$ or a surjective function $Y \to X$. You can do this either by using Zorn's Lemma (and mimicking the proof of Proposition 1.3.8), or by applying that proposition directly.

It follows from Proposition 1.3.8 that we can define the maximum of two cardinal numbers: $\max(|X|, |Y|) = |X|$ if $|Y| \leq |X|$, and it is $|Y|$ if $|X| \leq |Y|$.

This allows us to state the following properties of the arithmetic of cardinal numbers, which generalize Proposition 1.1.4. The proof makes essential use of Zorn's Lemma.

Proposition 1.3.9 *Let A and B be infinite sets. Then the following hold:*

  (i) $|A| + |A| = |A|$.
 (ii) $|A| + |B| = \max(|A|, |B|)$.
(iii) $|A| \times |A| = |A|$.
 (iv) $|A|^{|A|} = 2^{|A|}$.

Proof. For (i), we have to show that there is a bijective function: $A + A \to A$. To this end, we consider the poset of pairs (X, f) where X is a subset of A and $f : X + X \to X$ is bijective. This is ordered by: $(X, f) \leq (Y, g)$ if $X \subseteq Y$ and f is the restriction of g to $X + X$.

If $\{(X_i, f_i) \mid i \in I\}$ is a chain in this poset, then there is a well-defined bijective function $f : X + X \to X$, where $X = \bigcup_{i \in I} X_i$ and f is such that f extends each f_i.

Therefore the poset under consideration satisfies the hypothesis of Zorn's Lemma (check this!) and so has a maximal element (X, f). We claim that for this (X, f), $A - X$ must be finite.

To prove this claim, we use Proposition 1.2.2: if $A - X$ is infinite, there is an injective function $\mathbb{N} \to A - X$. Let $N \subseteq A$ be the image of this function; then we have a bijective function $g : N + N \to N$ (by Proposition 1.1.4), and since N and X are disjoint, we can combine f and g to obtain a bijective function

$$(X \cup N) + (X \cup N) \to X \cup N$$

which extends f; but this contradicts the maximality of the pair (X, f) in our poset. Therefore, $A - X$ is finite and we have proved the claim.

Now $A \sim X + (A - X)$, so by Exercise 19, there is a bijection $\varphi : A \to X$. Combining f and φ we obtain a bijection between $A + A$ and A:

$$A + A \overset{\varphi + \varphi}{\to} X + X \overset{f}{\to} X \overset{\psi}{\to} A$$

where ψ is the inverse of φ.

For (ii): suppose that $|A| \leq |B|$. We have to show that $|A| + |B| = |B|$. Since obviously $|B| \leq |A| + |B|$ and $|A| + |B| \leq |B| + |B|$ by hypothesis, using (i) we have

$$|B| \leq |A| + |B| \leq |B| + |B| \leq |B|$$

so $|B| = |A| + |B|$ as required. Note that this proof does not require that A is infinite.

For (iii), we form again a poset P of pairs (X, f) with $X \subseteq A$, but now with X infinite and $f : X \times X \to X$ bijective. By Propositions 1.2.2 and 1.1.4 (ii), the poset P is nonempty. We order this 'by extension' as in the proof of (i) (note that $X \subseteq Y$ implies $X \times X \subseteq Y \times Y$). In the same way as in (i) we see that every chain in P has an upper bound (check for yourself that if $\{X_i \mid i \in I\}$ is a chain of sets under the inclusion ordering, and $X = \bigcup_{i \in I} X_i$, then $X \times X = \bigcup_{i \in I} (X_i \times X_i)$).

By Zorn's Lemma, P has a maximal element (M, f). If $|M| = |A|$ we use the bijection between M and A, together with f, to obtain a bijection between $A \times A$ and A and we are done.

So suppose $|M| < |A|$. Now M is infinite by definition of P, and also $A - M$ is infinite: if $A - M$ were finite (hence countable), then by Exercise 19 we would have $M + (A - M) \sim M$ and therefore $A \sim M$, contradicting our assumption. Therefore, we can apply part (ii) to conclude that $A \sim A - M$.

Let $g : A \to A - M$ be bijective. If M' is the image of M under g, then M and M' are disjoint and g restricts to a bijection between M and M'. Combining g and

f, we also find a bijection $f' : M' \times M' \to M'$. Moreover, since M and M' are disjoint, we have $M \cup M' \sim M + M'$, and we can find a bijective function

$$F : (M \cup M') \times (M \cup M') \to M \cup M'$$

which extends f, as follows: we have

$$(M + M') \times (M + M') \sim (M \times M) + (M \times M') + (M' \times M) + (M' \times M').$$

We have $M \times M \sim M$ via f, and we have $(M \times M') + (M' \times M) + (M' \times M') \sim M'$ by using f, g, f' and part (i) of the proposition twice.

Finally, for (iv) we first notice that since A is infinite, $2 < |A|$ so $2^{|A|} \leq |A|^{|A|}$; for the converse inequality, we know from Proposition 1.1.5 that $|A| < 2^{|A|}$, so $|A|^{|A|} \leq (2^{|A|})^{|A|}$. Using (iii) of the proposition and Exercise 7, we see that $(2^{|A|})^{|A|} = 2^{|A| \times |A|} = 2^{|A|}$, and we are done.   ∎

Exercise 26

(a) Let A and B be nonempty sets, at least one of them infinite. Show that

$$|A| \times |B| = \max(|A|, |B|).$$

(b) Show that if A is infinite, then there is a bijection between A and $\mathbb{N} \times A$.
(c) Let A be an infinite set. Denote by A^* the set of all finite sequences of elements of A; that is,

$$A^* = \bigcup_{n=0}^{\infty} A^n$$

(here A^n is the set of n-tuples from A).
    Show that $|A^*| = |A|$.
(d) Let A be infinite; show that $|A| = |\mathcal{P}_{\text{fin}}(A)|$ (where $\mathcal{P}_{\text{fin}}(A)$ is the set of finite subsets of A).

Example 1.3.10 Let us come back to Example 1.3.6 and show that if both B and B' are bases of a vector space V, then $|B| = |B'|$. In other words, the vector space V has a well-defined *dimension* which is the cardinal number $|B|$ for a basis B of V. This fact *cannot* be proved (in full generality) without the Axiom of Choice!

We make use of the fact that a basis of a vector space V is a subset B which generates V (every element of V can be written as a finite linear combination of elements of B) but is *minimal* with this property: no proper subset of B generates V.

So, let B and B' be bases of V. We assume that you know the result in the case when B and B' are finite. So suppose B is infinite.

For every $b \in B'$ there is a finite subset B_b of B such that b can be written as a linear combination of elements of B_b. Then $\bigcup_{b \in B'} B_b$ is a subset of B which generates V, so by minimality of B,

$$B = \bigcup_{b \in B'} B_b.$$

It follows that B' is also infinite (otherwise, B would be a finite union of finite sets). Since every B_b is finite, there are injective functions $B_b \to \mathbb{N}$ and we see that

$$|B| \le |B'| \times \omega = |B'|.$$

By symmetry, $|B| = |B'|$, as required.

We close this section with some miscellaneous exercises involving Zorn's Lemma and cardinalities.

Exercise 27 Let X be an infinite set. Prove that there is a bijection $f : X \to X$ with the property that for every $x \in X$ and all $n > 0$, $f^n(x) \ne x$, where $f^n(x) = \underbrace{f(\cdots(f(x))\cdots)}_{n \text{ times}}$. [Hint: consider $\mathbb{Z} \times X$, or use Zorn directly.]

Exercise 28 Prove that there is a linear order on any set.

Exercise 29 Prove that there is a dense linear order on any infinite set: that is, a linear order such that whenever $x < y$, there is a z such that $x < z < y$. [Hint: use the previous exercise to find a linear order on X; then consider $\mathbb{Q} \times X$.]

Exercise 30 This exercise is one of the first applications, given by Cantor [8], of his theory of cardinalities to number theory.

A real number r is called *algebraic* if there is a non-zero polynomial

$$f(X) = X^n + a_1 X^{n-1} + \cdots + a_{n-1} X + a_n$$

with $a_1, \ldots, a_n \in \mathbb{Q}$ and $f(r) = 0$. A number which is not algebraic is called *transcendental*. Write A for the set of algebraic real numbers, and T for the set of transcendental real numbers.

Prove that $|A| = \omega$ and $|T| = |\mathbb{R}|$.

This was Cantor's proof that transcendental numbers exist, and that there are very many of them.

Exercise 31 In this exercise we consider $\mathbb{R}$ as a group under addition.

(a) Prove, using Zorn's Lemma, that there is a subgroup G of $\mathbb{R}$ which is maximal w.r.t. the property that $1 \notin G$.
(b) Suppose G is as in (a). Show that there is a unique prime number p such that $p \in G$.
(c) Let p be as in (b). Prove that for every $x \in \mathbb{R}$ there is an $n \ge 0$ such that $p^n x \in G$.

Exercise 32 This is a variation on the theme of Exercise 31. We consider the half-open interval $[0, 2) \subset \mathbb{R}$ as an additive group with addition modulo 2. As in Exercise 31, there is a subgroup G of $[0, 2)$ which is maximal w.r.t. the property that $1 \notin G$.

(a) Show that for each real number $x \in [0, 2)$ there is an even natural number n such that $nx \in G$.
(b) Show that for each $x \in [0, 2)$ the least natural number n satisfying $nx \in G$ is a power of 2.
(c) Show that for each odd natural number p we have $\frac{2}{p} \in G$. [Hint: first show that it is enough that for some power 2^k with $k > 0$, $2^k \frac{2}{p} \equiv \frac{2}{p} \pmod{2}$. Then use a little algebra to prove this fact.]

Exercise 33 For a subset A of $\mathbb{R}$ such that $0 \notin A$, we define:

$$\sqrt{\mathbb{Q}} = \{x \in \mathbb{R} \mid x^2 \in \mathbb{Q}\}$$
$$\mathbb{Q}/A = \{\tfrac{q}{a} \mid q \in \mathbb{Q} - \{0\}, a \in A\}.$$

Prove that there is a subset $A \subseteq \mathbb{R} - \{0\}$ such that $\mathbb{R}$ can be written as a *disjoint* union

$$A \cup \sqrt{\mathbb{Q}} \cup \mathbb{Q}/A.$$

[Hint: apply Zorn's Lemma to the poset of those $A \subseteq \mathbb{R}$ for which the following holds: for all $x, y \in A$, $xy \notin \mathbb{Q}$.]

Exercise 34 Let $(P, \leq)$ be a poset. Call a subset A of P *transversal* if for any pair x, y of elements of A we have: if $x \leq y$ then $x = y$. A subset B of P is called a *cross-section* of P if for every $x \in P$ there is a $y \in B$ satisfying $x \leq y$ or $y \leq x$.

(a) Prove that P has a maximal transversal subset.
(b) Prove that every maximal transversal subset is a cross-section.
(c) Prove that if A is a maximal transversal subset of P and $B \subsetneq A$ then B is not a cross-section of P.

Exercise 35

(a) Prove that there is a subset A of $\mathbb{R}$ which is maximal with respect to the properties:

   (i) A is closed under addition (i.e., whenever $a, b \in A$ then also $a + b \in A$).
   (ii) $A \cap \mathbb{Q} = \emptyset$.

(b) Let A be as in part (a). Prove that for every $x \in \mathbb{R}$ exactly one of the following (mutually exclusive) statements holds: $x \in A$, $x \in \mathbb{Q}$ or there is a natural number $m > 0$ and an $a \in A$ such that $mx + a \in \mathbb{Z}$. [Hint: given $x \in \mathbb{R}$ with $x \notin A$, $x \notin \mathbb{Q}$, consider the set

$$A_x = \{nx + ka \mid a \in A, n, k \in \mathbb{N}, n + k > 0\}.]$$

1.4 Well-Ordered Sets

Definition 1.4.1 A partial order $(L, \leq)$ is a *well-order*, or a *well-ordered set*, if every nonempty subset S of L has a *least element* in this order: there is an element $s_0 \in S$ such that for each $s \in S$, $s_0 \leq s$.
We shall also sometimes say that the relation $\leq$ *well-orders* L.

Recall that a partial order $(L, \leq)$ is linear or total if for all $x, y \in L$ we have $x \leq y$ or $y \leq x$.

Exercise 36 Prove that every well-order is linear.

Let us see some examples of well-orders.

(1) The set $\mathbb{N}$ is a well-ordered set. That this is so is exactly the principle of *induction* for natural numbers. We shall see later that conversely for every well-order there is a similar "induction principle" (Proposition 1.4.4).

(2) $\mathbb{Z}$ is not a well-ordered set (with the usual ordering): $\mathbb{Z}$ itself has no least element. In the same way, $\mathbb{Q}$ and $\mathbb{R}$ are not well-ordered in the usual ordering.

(3) Define a new ordering on $\mathbb{N}$ by putting: $n \leq m$ if either n and m are both odd and n is smaller than m in the usual ordering, or n and m are both even and n is smaller than m in the usual ordering, or n is even and m is odd. This looks like:

$$0 \leq 2 \leq 4 \leq \cdots \leq 1 \leq 3 \leq 5 \leq \cdots$$

and $\leq$ is a well-ordering. In a similar way, we can have:

$$0 \leq 3 \leq 6 \leq \cdots \leq 1 \leq 4 \leq 7 \leq \cdots \leq 2 \leq 5 \leq 8 \leq \cdots$$

and so on.

(4) Every finite linear order is a well-order.

(5) The set $\{1 - \frac{1}{n} \mid n > 0\} \cup \{1\}$ is a well-ordered subset of $\mathbb{R}$.

Exercise 37 Prove the following statements.

(a) If $(L, \leq)$ is a well-order, then so is every subset of L when equipped with the restricted order.

(b) If $(L, \leq_L)$ and $(M, \leq_M)$ are two well-ordered sets, we can define the *(reverse) lexicographic order* on the product $L \times M$: for elements (x, y) and (x', y') of $L \times M$, put $(x, y) \leq (x', y')$ if either $y <_M y'$, or $y = y'$ and $x \leq_L x'$. Then $\leq$ well-orders $L \times M$. The reverse lexicographic order can be pictured as follows: view M as points on a line (since M is linearly ordered) and replace every point of M by a copy of L.

Generalize this construction as follows: if L is a well-order and for each $i \in L$ we are given a well-order M_i, then there is a well-order on the set $\coprod_{i \in L} M_i$.

(c) This is a special case of the generalization in part (b):

   If $(L, \leq_L)$ and $(M, \leq_M)$ are two disjoint well-ordered sets, we can define an order on the union $L \cup M$ in the following way. For $x, y \in L \cup M$ we put $x \leq y$ if and only if one of the following three conditions is satisfied:

   (i) Both x and y are elements of L and $x \leq_L y$.
   (ii) We have $x \in L$ and $y \in M$.
   (iii) Both x and y are elements of M and $x \leq_M y$.

   Then $\leq$ well-orders $L \cup M$. This well-order on $L \cup M$, which we denote by $L + M$, looks like putting M "on top of" L.

In this exercise we defined the *reverse* lexicographic order, rather than the lexicographic order you might expect, for the sole reason that it corresponds to the multiplication of ordinals that we shall define in Chapter 4 (see Exercise 139).

The following proposition gives a criterion for whether or not a linear order is a well-order.

Proposition 1.4.2 *A linear order* $(L, \leq)$ *is a well-order if and only if for every infinite decreasing sequence*

$$x_0 \geq x_1 \geq x_2 \geq \cdots$$

in L, *there is an* n *such that for all* $m > n$, $x_n = x_m$.

Exercise 38 Prove Proposition 1.4.2. In one direction, you need to use the Axiom of Choice, in a way similar to the proof of Proposition 1.2.2.

Exercise 39 Recall that if $(X, \leq)$ is a poset, then $(X, \geq)$ is a poset too.
   Prove: if $(X, \leq)$ and $(X, \geq)$ are both well-orders, then X is finite.

We introduce some terminology for elements of a well-ordered set L. The empty set is vacuously well-ordered; it has no nonempty subsets.

   If L is nonempty, the set L (as nonempty subset of itself) has a least element, which we may as well call 0_L. If an element x is not the greatest element of L then the set $\{y \in L \mid x < y\}$ has a least element, which we call $x + 1$. So if L is infinite, L contains

$$\{0_L, 0_L + 1 = 1_L, 1_L + 1 = 2_L, 3_L, 4_L, \ldots\}$$

as a subset. If this subset is not the whole of L then its complement has a least element ω_L, and we may have $\omega_L, \omega_L + 1, \ldots$. This process may continue indefinitely!

   An element x of L is called a *successor* element if for some $y \in L$ we have $x = y + 1$. Elements which are not successors are called *limit* elements. Note, that 0_L is a limit element, as is ω_L.

Exercise 40 Let x and y be non-maximal elements of a well-order L. Show that if $x + 1 = y + 1$ then $x = y$.

In a well-order (or more generally, in a poset), a *least upper bound* or l.u.b. of a subset S is an upper bound x for S (see Definition 1.3.2) such that for every upper bound y for S we have $x \le y$. Note that least upper bounds need not always exist in a poset, but if they exist they are unique. Also note that x is a l.u.b. of the empty set if and only if x is the least element.

Proposition 1.4.3 *In a well-order $(L, \le)$ every subset of L which has an upper bound in L has a least upper bound. Moreover, an element x of L is a limit element if and only if x is the least upper bound of the set*

$$L_{<x} = \{y \in L \mid y < x\}.$$

Proof. If S has an upper bound in L, the set of upper bounds (in L) of S is nonempty, so it has a least element.

Suppose x is the l.u.b. of $L_{<x} = \{y \in L \mid y < x\}$. Then if $x = z + 1$ we must have that z is the greatest element of $L_{<x}$ and therefore its least upper bound; but z and $x = z + 1$ are distinct. We conclude that x is a limit element.

Conversely, suppose x is a limit element. Then for each $y < x$, x is not equal to $y + 1$, so $y + 1 < x$. It follows that no element smaller than x can be the l.u.b. of $L_{<x}$; but x is an upper bound for $L_{<x}$, so it is the least upper bound. ∎

We are now going to look at the principle of *induction* for well-ordered sets L. The well-known induction principle for natural numbers,

(I_0) If $S \subseteq \mathbb{N}$ has the properties that $0 \in S$ and for all $n \in \mathbb{N}$, $n \in S$ implies $n + 1 \in S$, then $S = \mathbb{N}$.

has an equivalent formulation:

(I_1) If $S \subseteq \mathbb{N}$ has the property that for each $n \in \mathbb{N}$, $n \in S$ whenever $\mathbb{N}_{<n} \subseteq S$, then $S = \mathbb{N}$.

In a similar way, we have two equivalent induction principles for an arbitrary well-ordered set L. We call a subset S of a well-order $(L, \le)$ *closed under least upper bounds* if whenever $A \subseteq S$ and A has an upper bound in L, then the least upper bound of A belongs to S.

Proposition 1.4.4 *Let $(L, \le)$ be a well-ordered set, and $S \subseteq L$ an arbitrary subset.*

(i) *If S is closed under the successor function (mapping x to $x + 1$) and under least upper bounds, then $S = L$.*
(ii) *If for each $x \in L$ the statement $L_{<x} \subseteq S$ implies $x \in S$, then $S = L$.*

Proof.

(i) If $S \ne L$ then $L - S$ has a least element x. Then we must have $\forall y < x \, (y \in S)$, yet $x \notin S$, which contradicts the assumption of (i).
(ii) is left to you as an exercise. ∎

Exercise 41 Prove Proposition 1.4.4 (ii). Prove also that the condition on S in Proposition 1.4.4 (ii) is equivalent to the following three conditions:

(a) $0_L \in S$.
(b) If $x \in S$, then $x + 1 \in S$.
(c) If l is a nonzero limit element of L and $L_{<l} \subseteq S$, then $l \in S$.

Example 1.4.5 Let us prove, by induction on L, that for each $x \in L$ there is a unique limit element $l \leq x$ and a unique natural number n such that

$$x = l + n$$

($l + n$ is shorthand for: the n-th successor of l, so $l + 0 = l$, etcetera).

Since the successor function is 1–1 where it is defined (Exercise 40), it is easy to show that such a representation is unique. Suppose $l + n = l' + n'$. Then by induction on n we show that $l = l'$ and $n = n'$: if $n = 0$, we have $l = l' + n'$, whence $n' = 0$ because l is a limit; and if $n = k + 1$ then $l + n$ is a successor, so $n' = k' + 1$ for some k'. By injectivity of the successor, we have $l + k = l' + k'$ and by induction hypothesis it follows that $l = l'$ and $k = k'$, so $n = n'$.
For the existence of the representation, we use induction on L: clearly, 0_L has the representation $0_L + 0$, for 0_L is a limit. If $x = l + n$ then $x + 1 = l + (n + 1)$, and if l is a limit, it has the representation $l + 0$.

For natural numbers, one has, beside induction for proving properties of natural numbers, also the possibility of defining functions by *recursion*: a function f is defined on natural numbers by a scheme which defines $f(n + 1)$ in terms of $f(n)$ or in terms of $\{f(k) \mid k \leq n\}$. An example is the well-known Fibonacci sequence: $f(0) = f(1) = 1$, and $f(n + 2) = f(n) + f(n + 1)$. Induction and recursion are really two sides of the same coin, so it is not surprising that we can also define functions on an arbitrary well-ordered set L by recursion. The idea is that one defines $f(x)$ in terms of the (not necessarily finite) *set* $\{f(y) \mid y < x\}$. There are various formulations. We prove one in the proposition below and give others as exercises.

Proposition 1.4.6 *Let* $(L, \leq)$ *be a well-order, and* S *a set. Suppose we are given a function* $R : L \times \mathcal{P}(S) \to S$. *Then there is a unique function* $F : L \to S$ *with the property that*

(∗) $F(l) = R(l, \{F(x) \mid x < l\})$

for each $l \in L$.
The function F *is said to be defined by* recursion *from* R.

Proof. In this proof, let $L_{\leq z}$ denote the set $\{y \in L \mid y \leq z\}$, for $z \in L$.
First, let us show that if F is a function from $L_{\leq z}$ to S such that F satisfies condition (∗), then F is unique with this property; for if

$$G : L_{\leq z} \to S$$

also satisfies $(*)$ and $F \neq G$, there must be a *least* element $m \in L_{\leq z}$ such that $F(m) \neq G(m)$; however in that case the sets $\{F(x) \mid x < m\}$ and $\{G(x) \mid x < m\}$ are equal so that by $(*)$, $F(m) = G(m)$, which contradicts our assumption on m.

Similarly, any $F : L \to S$ satisfying $(*)$ must be unique.

From this uniqueness it follows that if $z_1 < z_2$ in L and $F_1 : L_{\leq z_1} \to S$ and $F_2 : L_{\leq z_2} \to S$ satisfy $(*)$, then F_1 must be the restriction of F_2 to the subset $L_{\leq z_1}$.

Therefore, if for each $z \in L$ a function $F_z : L_{\leq z} \to S$ exists which satisfies $(*)$, the functions F_z can be patched together to a unique function

$$F : L = \bigcup_{z \in L} L_{\leq z} \to S$$

and F also satisfies $(*)$ because for $z \in L$ we have $F(z) = F_z(z)$.

We see that in order to finish the proof it is enough to show that for each $z \in L$, a function F_z as above exists. We do this by induction on L: for an application of Proposition 1.4.4 (i) let E be the set $\{z \in L \mid F_z \text{ exists}\}$. We wish to show $E = L$.

Suppose, for a given $z \in L$, that $w \in E$ for all $w < z$; that is, $F_w : L_{\leq w} \to S$ exists and satisfies $(*)$. We define $F_z : L_{\leq z} \to S$ by putting

$$F_z(w) = F_w(w) \qquad \text{for } w < z$$
$$F_z(z) = R(z, \{F_w(w) \mid w < z\}).$$

Check yourself that F_z satisfies $(*)$. We have proved that $z \in E$ on the assumption that $w \in E$ for all $w < z$. In other words, the subset E satisfies the hypothesis of Proposition 1.4.4 (i) and we may conclude that $E = L$ as desired. ■

Exercise 42 Let $(L, \leq)$ be a well-order and S a set. Prove the following two variations on the principle of recursion.

(i) For any $R : \mathcal{P}(S \times L) \to S$ there is a unique $F : L \to S$ with

$$F(x) = R(\{(F(y), y) \mid y < x\}).$$

(ii) For any $s_0 \in S$, any $R : \mathcal{P}(S) \to S$ and any $g : S \to S$, there is a unique function $F : L \to S$ which satisfies the following conditions:

$$F(0_L) = s_0.$$
$$F(l + 1) = g(F(l)) \qquad \text{if } l \text{ is not maximal in } L.$$
$$F(l) = R(\{F(y) \mid y < l\}) \text{ if } l \text{ is a nonzero limit in } L.$$

Example 1.4.7 Let us give a simple example of a function $L \to \{0, 1\}$ defined by recursion: the *parity* function. Consider the formulation of Exercise 42 (ii) with $S = \{0, 1\}$, $s_0 = 0$, g the switch function $(g(x) = 1 - x)$ and R the constant 0 function. One obtains a unique function $F : L \to \{0, 1\}$ such that $F(x)$ is $n \bmod 2$, where n is the unique natural number such that for some limit element l, $x = l + n$.

More fundamental examples of functions defined by recursion appear in the proof of Proposition 1.4.11 below, and in Section 1.5.

We conclude this section by discussing how to compare well-orders.

Definition 1.4.8 Let $(L, \leq)$ and $(M, \leq)$ be well-orders.

(i) An *initial segment* of L is a subset $B \subseteq L$ such that for each $x, y \in L$: if $x \in B$ and $y \leq x$, then $y \in B$.
(ii) An *(order-)isomorphism* $f : L \to M$ is an order-preserving bijective function.
(iii) An *embedding* $f : L \to M$ is an order-isomorphism from L to an initial segment of M.

We write $L \cong M$ and say that L is *isomorphic to* M if there is an order-isomorphism between L and M. We write $L \preceq M$ if there exists an embedding of L into M.

Lemma 1.4.9 *There can be at most one embedding from a well-order L into a well-order M. If L is a well-ordered set and $l \in L$ there cannot be an embedding $L \to L_{<l}$. In particular, L cannot be isomorphic to $L_{<l}$.*

Proof. Suppose f and f' are two different embeddings: $L \to M$. Then the set $\{x \in L \mid f(x) \neq f'(x)\}$ is nonempty and has a least element x_0. We may suppose (since M is in particular a total order) that $f(x_0) < f'(x_0)$. But now we have: if $y < x_0$ then $f'(y) = f(y) < f(x_0)$ and if $y \geq x_0$ then $f'(y) \geq f'(x_0) > f(x_0)$. We conclude that $f(x_0)$ is not in the image of f', which is therefore not an initial segment.

For the second statement we notice that the inclusion of $L_{<l}$ into L is an embedding. Given an embedding $f : L \to L_{<l}$, we can compose f with this inclusion map to get an embedding $L \to L$, which is not the identity since its image does not contain the element l. Hence we would have at least two different embeddings $L \to L$, contradicting the first statement.

For the third statement, note that every isomorphism is an embedding. ∎

Corollary 1.4.10 *If $L \preceq M$ and $M \preceq L$ then $L \cong M$.*

Proof. If $i : L \to M$ and $j : M \to L$ are embeddings, then the composition $ji : L \to L$ is an embedding too. Since there is only one embedding from L to L by Lemma 1.4.9 and the identity function $f(x) = x$ is one, we see that $j(i(x)) = x$ for all $x \in L$. Similarly, $i(j(y)) = y$ for all $y \in M$; so $L \cong M$. ∎

Proposition 1.4.11 *For any two well-orders L and M, we have either $L \preceq M$ or $M \preceq L$.*

Proof. Let ∞ be a new point not contained in M. Let $M' = M \cup \{\infty\}$.

Define $R : L \times \mathcal{P}(M') \to M'$ as follows: $R(l, S)$ is the least element in M of $M - S$, if this set is nonempty, and ∞ otherwise. The function R does not really depend on l, but that doesn't matter.

By Proposition 1.4.6 there is a unique function $F : L \to M'$ such that

$$F(l) = R(l, \{F(x) \mid x < l\})$$

for all $l \in L$.

If $F(l) \neq \infty$ for all $l \in L$, then F is an embedding from L into M (as we leave for you to check). Otherwise, if l_0 is the least element of L such that $F(l_0) = \infty$, then F restricts to an isomorphism between M and $L_{<l_0}$, in which case there is an embedding of M into L.   ∎

Exercise 43   Prove: if L and M are well-orders such that $L \preceq M$ and $L \not\cong M$, then there is an $m \in M$ such that $L \cong M_{<m}$ where $M_{<m} = \{x \in M \mid x < m\}$. Prove also that this m is unique.

Exercise 44   Let L be a well-order and $f : L \to L$ a map with the property that $x < y$ implies $f(x) < f(y)$ for all $x, y \in L$.

Show that $x \leq f(x)$ for all $x \in L$. Show also that this does not follow from the weaker condition that $x \leq y$ implies $f(x) \leq f(y)$.

Exercise 45   Let L be a set. Write $\mathcal{P}^*(L)$ for the set of nonempty subsets of L. Suppose that $h : \mathcal{P}^*(L) \to L$ is a function such that the following two conditions are satisfied:

(i) For each nonempty family $\{A_i \mid i \in I\}$ of elements of $\mathcal{P}^*(L)$, we have

$$h(\bigcup_{i \in I} A_i) = h(\{h(A_i) \mid i \in I\}).$$

(ii) For each $A \in \mathcal{P}^*(L)$, $h(A) \in A$.

Show that there is a unique relation $\leq$ on L which well-orders L, and is such that for each nonempty subset A of L the element $h(A)$ is the least element of A.

Exercise 46   Let L be a linear order. If $A \subset L$ and $a \in L$, then a is called a *strict upper bound* for A if $x < a$ for every $x \in A$. Now suppose that the following is true for every $A \subseteq L$: if A has a strict upper bound then A has a *least* strict upper bound.

(a) Prove: if $L \neq \emptyset$, then L has a least element.
(b) Prove that L is a well-order. [Hint: given a nonempty subset X of L, consider the set $A_X = \{x \in L \mid \text{for all } y \in X, x < y\}$.]
(c) Show that (b) may fail if we drop the 'strict' in 'strict upper bound'.

Exercise 47   Extend the definition of 'initial segment' (Definition 1.4.8) to arbitrary linear orders: an initial segment of $(P, \leq)$ is a subset $B \subseteq P$ such that $x \in B$ whenever $x \leq y$ and $y \in B$.

Prove that a linear order $(P, \leq)$ is a well-order if and only if for every subset S of P the linear order $(S, \leq)$ is isomorphic to an initial segment of $(P, \leq)$.

Exercise 48   A poset $(W, \leq)$ is called *well-founded* if every nonempty subset Y of W has a *minimal element* y_0 (i.e., for every $y \in Y$ we have: if $y \leq y_0$, then $y = y_0$).

(a) Prove that $(W, \leq)$ is well-founded if and only if W contains no infinite strictly decreasing sequence $x_0 > x_1 > \cdots$. Compare with Proposition 1.4.2.
(b) Conclude from (a) that a well-founded poset is a well-order if and only if it is linear.
(c) Prove the *principle of well-founded induction*: if the poset $(W, \leq)$ is well-founded and a subset $S \subseteq W$ has the property that $W_{<x} \subseteq S$ implies $x \in S$ for any $x \in W$, then $S = W$. Compare with Proposition 1.4.4.
(d) Prove that Proposition 1.4.6 also holds for well-founded posets.
(e) Prove that a poset $(W, \leq)$ is well-founded if and only if there exists a function from W into a well-order, which preserves the strict order $<$.

1.5   Principles Equivalent to the Axiom of Choice

We have seen the Axiom of Choice, Zorn's Lemma and the Principle of Cardinal Comparability. In this section, we formulate another principle of importance, Zermelo's *Well-Ordering Theorem* (Definition 1.5.2), and prove rigorously that these four principles are equivalent.

In fact, these are just a few examples out of many: a multitude of statements have been shown to be equivalent to the Axiom of Choice (you may have a look at the books [38, 66]). Here we mention just two more, without proof:

Hausdorff's Maximality Principle (Felix Hausdorff, 1868–1942) says that every poset contains a maximal chain (maximal w.r.t. inclusion of chains). It is actually rather easy to show that this is equivalent to Zorn's lemma.

Tychonoff's Theorem (Andrej Nikolayevich Tychonoff, 1906–1993) in Topology says that if $\{X_i \mid i \in I\}$ is an arbitrary set of compact topological spaces then the product space

$$\prod_{i \in I} X_i$$

is again a compact space. The proof of Tychonoff's Theorem makes use of AC. On the other hand, it can be shown that the theorem implies AC (a fact first proved in [42]; see also [38]).

The following lemma ensures the existence of "sufficiently large" well-ordered sets. It should be emphasized that its proof does *not* use the Axiom of Choice or Zorn's Lemma. The lemma was published by Friedrich Hartogs (1874–1943) in [31].

Lemma 1.5.1 (Hartogs' Lemma)   *For every set X there is a well-ordering $(L_X, \leq)$ such that there is no injective function from L_X to X.*

Proof. Let $\mathcal{P}$ be the set of all pairs $(L, \leq)$ where L is a subset of X and $\leq$ is a well-ordering on L. We shall denote the pair $(L, \leq)$ simply by L.

For two such L and M we write as usual $L \preceq M$ if there is a (necessarily unique, by Lemma 1.4.9) embedding of well-ordered sets: $L \to M$. We note the following facts.

We have both $L \preceq M$ and $M \preceq L$ if and only if $L \cong M$.

If $L \cong L'$ and $M \cong M'$, then $L \preceq M$ if and only if $L' \preceq M'$.

Therefore the relation $\preceq$ is defined on the set Q of equivalence classes of P modulo the equivalence relation $\cong$. By Proposition 1.4.11, the set Q is a linear order with the relation $\preceq$.

Note that, if $L \prec M$ (that is, $L \preceq M$ but $M \not\preceq L$), there is (by Exercise 43) a unique $m \in M$ such that L is isomorphic to the set $M_{<m} = \{m' \in M \mid m' < m\}$ with the inherited order from M.

Therefore, if we denote the $\cong$-equivalence class of L by $[L]$, the set

$$\{\alpha \in Q \mid \alpha < [L]\}$$

is isomorphic to L.

Now suppose that $W \subseteq Q$ is a nonempty set of $\cong$-equivalence classes. Let $\alpha = [L]$ be an arbitrary element of W. Consider the set

$$L_W = \{l \in L \mid [L_{<l}] \in W\}.$$

If L_W is empty, clearly $[L]$ is the least element of W. If L_W is nonempty then it has (as a subset of the well-ordered set L) a least element l_W. But then $[L_{<l_W}]$ is the least element of W. So every nonempty subset of Q has a least element, and therefore Q is a well-ordered set.

There cannot be an injective function from Q into X, for suppose f is such a function. Then f gives a bijection from Q to a subset U of X and we can give U the induced well-ordering of Q; denote this well-ordering by $\leq_U$. Now $(U, \leq_U)$ represents an element of Q and for every $u \in U$ we have

$$[(U_{<u}, \leq_U)] < [(U, \leq_U)]$$

in Q, by Lemma 1.4.9. If we now define a function $g : Q \to Q$ by

$$g(\alpha) = [(U_{<f(\alpha)}, \leq_U)]$$

then g is an embedding $Q \to Q_{<[(U,\leq_U)]}$, but this contradicts Lemma 1.4.9. ∎

In his paper [85], Zermelo formulated the Axiom of Choice in order to prove the following statement.

Definition 1.5.2 The *Well-Ordering Theorem* is the statement that for every set X there exists a relation $\leq$ which well-orders X.

Remark 1.5.3 Although the Axiom of Choice is intuitively correct, here we see a consequence which is less intuitive, for it asserts that there is a relation which well-orders the set of real numbers, for example. However, it can be shown that it is impossible to define such a relation explicitly ([19]).

Proposition 1.5.4 *The following assertions are equivalent:*

  (i) *The Axiom of Choice.*
 (ii) *Zorn's Lemma.*
(iii) *The Principle of Cardinal Comparability.*
(iv) *The Well-Ordering Theorem.*

Proof. We shall prove (i)$\Rightarrow$(ii)$\Rightarrow$(iii)$\Rightarrow$(iv)$\Rightarrow$(i).
The implication (i)$\Rightarrow$(ii) uses Hartogs' Lemma (1.5.1). Suppose that $(P, \leq)$ is a poset in which every chain has an upper bound, yet P has no maximal element. We shall prove, using the Axiom of Choice, that in that case for every well-ordered set L there is an embedding of L into P.

Since in P every chain has an upper bound, P is nonempty; let $p_0 \in P$. By the Axiom of Choice there is a function $R : \mathcal{P}(P) \to P$ such that for every subset C of P we have: if C is a chain in P then $R(C)$ is an upper bound for C; and if not, $R(C) = p_0$. Also, since P has no maximal element, for every $p \in P$ there is a $q \in P$ with $p < q$; again using AC, there is a function $g : P \to P$ such that $p < g(p)$ for every $p \in P$.

Let $(L, \leq)$ be an arbitrary well-ordered set. Define a function $F : L \to P$ by recursion over L as follows:

- $F(0_L) = p_0$.
- $F(x + 1) = g(F(x))$.
- $F(l) = R(\{F(x) \mid x < l\})$ if l is a non-zero limit element of L.

It is easy to check that F is an injective function from L into P. L was arbitrary, so we get a contradiction with Hartogs' Lemma.
The implication (ii)$\Rightarrow$(iii) is Proposition 1.3.8.
The implication (iii)$\Rightarrow$(iv) uses Hartogs' Lemma once again. Let X be a set. According to Hartogs' Lemma there is a well-ordered set $(L_X, \leq)$ such that there is no injective function from L_X into X. By Cardinal Comparability then, there must be an injective function from X into L_X; but this gives us a well-ordering on X.
Finally, (iv)$\Rightarrow$(i) is easy. Suppose $f : X \to Y$ is surjective. In order to find a section for f, apply iv) to find a relation $\leq$ on X which well-orders X. Now one can simply define a section $s : Y \to X$ by letting $s(y)$ be the least element of the nonempty set $f^{-1}(y)$ in the well-ordering on X.   ∎

Exercise 49 Give a direct proof of the fact that Zorn's Lemma implies the Well-Ordering Theorem.

Exercise 50 Let X be a set, and S a subset of $\mathcal{P}(X)$. We say that S is *of finite character* if for every $A \subseteq X$ it holds that A is an element of S if and only if every finite subset of A is an element of S.

The *Teichmüller–Tukey Lemma* (Oswald Teichmüller, 1913–1943; John Tukey, 1915–2000) states that if S is nonempty and of finite character, S contains a maximal element (with respect to the subset ordering).

(a) Use Zorn's Lemma to prove the Teichmüller–Tukey Lemma.
(b) Show that the Teichmüller–Tukey Lemma implies the Axiom of Choice.

Exercise 51 According to Hartogs' Lemma, applied to the set $\mathbb{N}$, there is a well-ordered set W which is uncountable.

(a) Deduce from this that there exists a well-ordering $(L, \leq)$ with the following properties:

   (i) L is uncountable.
   (ii) For every $x \in L$, the set $L_{<x}$ is countable.

   [Hint: if W is an uncountable well-order, consider the set of those elements x of W for which the set $W_{<x}$ is uncountable. Alternatively, one may show that the statement follows directly from the proof of Hartogs' Lemma.]
(b) Let L be as in (a). Prove that if $x_0 < x_1 < x_2 < \cdots$ is an increasing sequence in L then this sequence has an upper bound in L.

Exercise 52 Let L and M be well-orders and suppose L has no greatest element.

(a) Prove: if for every $l \in L$ there is an embedding $L_{<l} \to M$ then there is an embedding $L \to M$.
(b) Does the statement in part (a) still hold if we replace "embedding" by "injective function" throughout? Justify your answer.

Chapter 2
Models

In this chapter we develop the notion of a "formal language" as promised in the Introduction; and also its "interpretation" in mathematical structures.

In the nineteenth century, a number of mathematicians started to reflect on Logic; that is to say, the reasoning principles that are used in mathematical arguments (before that time, Logic belonged to the realm of Philosophy and consisted in studying *syllogisms* – separate reasoning steps – such as had been formulated by Aristotle).

It occurred to George Boole (1815–1864) and Augustus de Morgan (1806–1871) that the mathematical use of the words "and", "or" and "not" obeys the rules of algebra. This is why we have "Boolean rings". Further steps, introducing quantifiers ("for all" and "there exists"), were taken by Charles Sanders Peirce (1839–1914), but the most important work of this era is *Begriffsschrift* of Friedrich Ludwig Gottlob Frege (1848–1925), which appeared in 1879. "Begriffsschrift" can be roughly translated as "the notation of concepts". Frege not only defined a complete logical language, but also set out to develop mathematics in it. He abruptly abandoned the whole project after Bertrand Russell (1872–1970) pointed out an antinomy in his work, but Russell himself continued it in *Principia Mathematica* with A.N. Whitehead (1861–1947).

By this time (around 1900), the developing field of Logic had captured the attention of leading mathematicians such as David Hilbert and Henri Poincaré.

The idea that abstract mathematical statements (and therefore also the "sentences" of a logical language) can be interpreted in various "models" certainly existed in the first decades of the twentieth century. In fact it is already implicit in the proof by Lobachevsky (1792–1856) from 1826 that the parallel postulate is independent of the other axioms of Euclidean geometry. However, the formal definition of the notion "sentence ϕ is true in model X" is usually attributed to Alfred Tarski (1901–1983): see [75] for the German translation of his original paper.

© Springer Nature Switzerland AG 2018
I. Moerdijk, J. van Oosten, *Sets, Models and Proofs*, Springer Undergraduate
Mathematics Series, https://doi.org/10.1007/978-3-319-92414-4_2

Alfred Tarski

Certainly, Tarski created *Model Theory*, of which you will get a first glimpse in this chapter.

2.1   Rings and Orders: Examples

This section is meant to serve as introduction and motivation for the formal definition of an abstract *language* in the next section.

When we say that the real numbers $\mathbb{R}$ form a *commutative ring with* 1, we mean that there are two *distinguished* elements 0 and 1, as well as *operations* $+$ and $\cdot$, such that certain *axioms* hold, for example:

$$x \cdot (y + z) = (x \cdot y) + (x \cdot z).$$

This is to be read as: whenever real numbers are substituted for the *variables* x, y and z, we get an equality as above.

We call the whole of $\{0, 1, +, \cdot\}$ the *ring structure* of $\mathbb{R}$. Now of course you know there are plenty of other rings. For example, let X be any set. The power set $\mathcal{P}(X)$ can be made into a commutative ring with 1: take X for 1, $\emptyset$ for 0, and let for $U, V \subseteq X$, $U + V = (U \cup V) - (U \cap V)$ and $U \cdot V = U \cap V$.

Exercise 53 Check that this indeed gives a ring structure on $\mathcal{P}(X)$.

The example of $\mathcal{P}(X)$ makes it clear that the operation $+$ does not, a priori, mean addition of numbers, but is an abstract symbol generally used for the operation in abelian groups; we might as well have used something like $a(x, y)$ and $m(x, y)$ instead of $x + y$ and $x \cdot y$, respectively, and written the distributivity axiom as

$$m(x, a(y, z)) = a(m(x, y), m(x, z)).$$

Similarly, one should regard 0 and 1 as abstract symbols that only acquire meaning once they are *interpreted* in a particular set.

Many axioms for rings have a very simple form: they are equalities between *terms*, where a term is an expression built up using variables, the symbols 0 and 1, and the operation symbols $+$, $\cdot$ (and brackets). From simple equalities we can form more involved statements using *logical operations*: the operations $\wedge$ ("and"), $\vee$ ("or"), $\rightarrow$ ("if...then"), $\leftrightarrow$ ("if and only if") and $\neg$ ("not"); and *quantifiers* $\exists$ ("there is") and $\forall$ ("for all"). For example, if we want to express that $\mathbb{R}$ is in fact a field, we may write

$$\forall x(\neg(x = 0) \rightarrow \exists y(x \cdot y = 1))$$

or equivalently

$$\forall x \exists y(x = 0 \vee x \cdot y = 1).$$

We say that the statement $\forall x \exists y(x = 0 \vee x \cdot y = 1)$ "is true in $\mathbb{R}$" (of course, what we really mean is: in $\mathbb{R}$ together with the meaning of 0, 1, $\mid$, $\cdot$). Such statements can be used to distinguish between various rings: for example, the statement

$$\exists x(x \cdot x = 1 + 1)$$

is true in $\mathbb{R}$ but not in the ring $\mathbb{Q}$, and the statement

$$\forall x(x \cdot x = x)$$

is true in the ring $\mathcal{P}(X)$ but not in $\mathbb{R}$.

Apart from operations on a set, one may also consider certain *relations*. In $\mathbb{R}$ we have the relation of *order*, expressed by $x < y$. As before, we might have used a different symbol for this relation, for example $L(x, y)$ ("x is less than y"). And we can form statements using this new symbol together with the old ones, for example

$$\forall x \forall y \forall z(x < y \rightarrow x + z < y + z)$$

which is one of the axioms for an *ordered ring*. In $\mathbb{R}$, the order relation is *definable* from the ring structure, because the statement

$$\forall x \forall y(x < y \leftrightarrow \exists z(\neg(z = 0) \wedge x + z \cdot z = y))$$

is true in $\mathbb{R}$. However, this statement is not true in the ordered ring $\mathbb{Q}$. Also the ring $\mathcal{P}(X)$ is (partially) ordered by $U \subset V$; in this ring, the order is definable, but now in a different way:

$$\forall x \forall y(x < y \leftrightarrow (\neg(x = y) \wedge x \cdot y = x)).$$

In yet another way, the order in $\mathbb{Q}$ is definable from the ring structure. In this case, we use the theorem (first proved by Lagrange) which says that every natural number may be written as the sum of four squares. Since every positive rational number is the quotient of two positive natural numbers, we have:

$$x > 0 \leftrightarrow \exists y_1 \cdots y_8 \, (x \cdot (y_1^2 + \cdots + y_4^2 + 1) = y_5^2 + \cdots + y_8^2 + 1)$$

for all $x \in \mathbb{Q}$. Since $x < y$ is equivalent to $\exists z(z > 0 \wedge x + z = y)$, we can define the order on $\mathbb{Q}$ in terms of 0, $+$ and $\cdot$ only.

We see that in general, when we wish to discuss a certain type of mathematical structure, we choose symbols for the distinguished elements, the operations and the relations which make up the structure, and using these we write down statements. The use of such statements is varied: they may be axioms, required to be true in all structures we wish to consider; they may be true in some, but not in others; or they may be used to *define* elements or subsets of a structure.

In Mathematical Logic we study these statements and their relation to mathematical structures *formally*. In order to do this *we define formal statements as mathematical objects*. This is done in the next section.

We shall see many examples of different types of structures in the coming sections.

2.2 Languages of First-Order Logic

This section is purely "linguistic" and introduces the formal languages for first-order logic – or "predicate logic".

Definition 2.2.1 A *language* L is given by three sets of symbols: *constants*, *function symbols* and *relation symbols*. We may write

$$L = (\mathrm{con}(L), \mathrm{fun}(L), \mathrm{rel}(L)).$$

Moreover, for each function symbol f and each relation symbol R a natural number n of arguments is specified, and called the *arity* of f (or R). If f or R has arity n, we say that it is an *n-ary* (or *n-place*) function (relation) symbol. Instead of 1-ary, 2-ary and 3-ary we also use the words *unary*, *binary*, *ternary*. Since we shall consider a 0-place function symbol to be a constant, the arity of a function symbol is always a number greater than 0; but we do allow 0-ary relation symbols (although we shall not see many of those in this book).

Example 2.2.2 The language of rings has two constants, 0 and 1, and two 2-place (binary) function symbols for addition and multiplication. There are no relation symbols. The language of orders has one binary relation symbol (S or $\leq$) for "less than or equal to".

Given such a language L, one can build *terms* (to denote elements) and *formulas* (to state properties). Note that we will first *define* what terms and formulas *are* (mathematically) in Definitions 2.2.3 and 2.2.7, and then *how we write them* in Definitions 2.2.5 and 2.2.9.

We use the following auxiliary symbols:

- A countably infinite set of *variables*. This set is usually left unspecified, and its elements are denoted by $x, y, z, \ldots$ or $x_0, x_1, \ldots$
- The equality symbol $=$
- The symbol $\perp$ ("absurdity")
- Connectives: the symbols $\wedge$ ("and") for *conjunction*, $\vee$ ("or") for *disjunction*, $\rightarrow$ ("if... then") for *implication* and $\neg$ ("not") for *negation*
- Quantifiers: the *universal quantifier* $\forall$ ("for all") and the *existential quantifier* $\exists$ ("there exists")

We assume that all auxiliary symbols are distinct, and that the set of auxiliary symbols, the sets $\mathrm{con}(L)$, $\mathrm{rel}(L)$, $\mathrm{fun}(L)$ and the set of variables are pairwise disjoint.

We will define the "terms" of a language L, making sure that for every constant of L we have a term, for every variable we have a term, and whenever f is an n-place function symbol of the language L and $t_1, \ldots, t_n$ are terms of L, we will have a term indicating the application of the "function" f to "elements" $t_1, \ldots, t_n$. A precise and formal treatment defines terms as strings of symbols.

Let $\mathcal{C}_L$ be the collection of all symbols: the symbols of L and the auxiliary symbols. Write $\mathcal{C}_L^*$ for the set of all finite strings of these symbols. Such a finite string is written $\langle a_0, \ldots, a_{n-1} \rangle$ where $a_0, \ldots, a_{n-1} \subset \mathcal{C}_L$. For $n = 0$ this is the empty string $\langle \rangle$. On $\mathcal{C}_L^*$ we have the operation of *concatenation*: given two strings $s = \langle a_0, \ldots, a_{n-1} \rangle$ and $t = \langle b_0, \ldots, b_{m-1} \rangle$ the concatenation $s * t$ is the string

$$\langle a_0, \ldots, a_{n-1}, b_0, \ldots, b_{m-1} \rangle.$$

Of course, we can have multiple concatenations and since this operation is associative we can simply write $s_1 * s_2 * \cdots * s_n$ for an iterated concatenation.

We shall now single out specific subsets of $\mathcal{C}_L^*$: the set of *terms* of the language L and the set of *formulas* of L.

Definition 2.2.3 The set of *terms* of a language L (or L-*terms*) is the smallest subset T of $\mathcal{C}_L^*$ which satisfies the following conditions:

(i) If c is a constant of L then $\langle c \rangle$ is an element of T.
(ii) If x is a variable then $\langle x \rangle \in T$.
(iii) If $t_1, \ldots, t_n$ is an n-tuple of elements of T and f is an n-place function symbol of L, then the string

$$\langle f \rangle * t_1 * \cdots * t_n$$

is an element of T.

It is easy to see that for any collection of subsets of C_L^* satisfying conditions (i)–(iii) of Definition 2.2.3 its intersection also satisfies (i)–(iii). Therefore there exists a smallest such subset, and Definition 2.2.3 is correct.

Example 2.2.4 Given a term of a language L, it is possible to recover uniquely the steps by which it was constructed. For example, consider the string

$$\langle f, g, x, h, x, c, h, y, d, d \rangle$$

where f is a 3-place function symbol, g and h are 2-place function symbols, c and d are constants, and x and y are variables. Working "backwards" we identify the "subterms" $\langle d \rangle$, $\langle h, y, d \rangle$ and $\langle h, x, c \rangle$ which we may write as d, $h(y, d)$ and $h(x, c)$, respectively; then, the subterm $\langle g, x, h, x, c \rangle$ which we may write as $g(x, h(x, c))$ and finally the given string, which we may denote by

$$f(g(x, h(x, c)), h(y, d), d).$$

This justifies the following notation, which will be used throughout in this book.

Definition 2.2.5 (Notation of terms) For a variable x, we will write just x for the term $\langle x \rangle$. Similarly, for a constant c we will just write c for the corresponding term. Given an n-place function symbol f and terms $t_1, \ldots t_n$ we write $f(t_1, \ldots, t_n)$ for the term constructed in clause (iii) of Definition 2.2.3.

A term which does not contain variables (and hence is built up from constants and function symbols alone) is called *closed*.

In our notation $f(t_1, \ldots, t_n)$ we have used brackets and separated the terms by commas, to make for easier reading. This is not necessary for parsing the term, as the following exercise shows (this generalizes the example above).

Exercise 54 Show that if $t \in C_L^*$ is a concatenation of terms, then we can recover these terms uniquely from t.

Examples.

(a) Suppose L has a constant c and a 2-place function symbol f. The following are terms of L: $x, y, c, f(x, c), f(f(x, c), c), \ldots$
(b) Suppose L has no function symbols. The only terms are variables and constants.

It is an immediate consequence of the way the set of terms was defined that there is an *induction principle* for terms: if T is any subset of C_L^* which satisfies conditions (i)–(iii) of Definition 2.2.3, then T contains the set of L-terms as a subset. The following definition and exercise form an example of this.

Definition 2.2.6 Let t and s be elements of C_L^* and let $x \in C_L$ be a symbol. The *substitution* $t[s/x]$ is defined by recursion on the length of the string t as follows. Suppose $t = \langle a_0, \ldots, a_{n-1} \rangle$. Then we define:

(i) If $n = 0$ then $t[s/x] = t = \langle \rangle$.

(ii) If $n > 0$ and $a_0 = x$ then $t[s/x] = s * (\langle a_1, \ldots, a_{n-1} \rangle [s/x])$.
(iii) If $n > 0$ and $a_0 \neq x$ then $t[s/x] = \langle a_0 \rangle * (\langle a_1, \ldots, a_{n-1} \rangle [s/x])$.

Exercise 55 Suppose that t and s are L-terms and x is a variable. Prove that $t[s/x]$ is an L-term. [Hint: first prove that if t is of the form $f(t_1, \ldots, t_n)$ then $t[s/x]$ is $f(t_1[s/x], \ldots, t_n[s/x])$. Then carry out induction on t as an L-term.]

For formulas we do almost the same thing. Just to give a few construction principles: if t and s are L-terms, we will have an L-formula which expresses the equality of these terms; if R is an n-place relation symbol of L and $t_1, \ldots, t_n$ are L-terms, we have an L-formula which expresses that the n-tuple of "elements" $(t_1, \ldots, t_n)$ is in the "relation" R; if φ and ψ are L-formulas, we will have a formula expressing "φ and ψ". The precise treatment also defines formulas as strings of symbols:

Definition 2.2.7 The set of *formulas* of a given language L (or L-formulas) is the smallest subset F of $\mathcal{C}_L^*$ which satisfies the following conditions:

(i) If t and s are L-terms then $\langle = \rangle * t * s$ is an element of F.
(ii) If $t_1, \ldots, t_n$ are L-terms and R is an n-place relation symbol of L, then

$$\langle R \rangle * t_1 * \cdots * t_n$$

is an element of F. Note that if R is 0-ary this means that $\langle R \rangle$ is an element of F.
(iii) The string $\langle \perp \rangle$ is an element of F.
   Formulas formed by (i), (ii) or (iii) are called *atomic formulas*.
(iv) If φ and ψ are elements of F then the following are elements of F:

$$\langle \wedge \rangle * \varphi * \psi$$
$$\langle \vee \rangle * \varphi * \psi$$
$$\langle \rightarrow \rangle * \varphi * \psi$$
$$\langle \neg \rangle * \varphi$$

(v) If φ is an element of F and x is a variable then the following are elements of F:

$$\langle \forall \rangle * \langle x \rangle * \varphi$$
$$\langle \exists \rangle * \langle x \rangle * \varphi$$

Just as in the case of L-terms there is a smallest subset of $\mathcal{C}_L^*$ satisfying conditions (i)–(v) of Definition 2.2.7. Accordingly, we have an induction principle for formulas analogous to the one for terms.

Exercise 56 Prove by induction on formulas that the last symbol of a formula is a variable or a constant, or $\perp$ or a 0-place relation symbol R.

Example 2.2.8 Let us indicate by example that, just as with terms, it is possible to recover the whole construction process of a formula φ. Consider the formula

$$\langle \exists, x, \rightarrow, R, x, y, \forall, z, \wedge, \neg, =, f, y, z, c, \exists, v, S, x, y, v \rangle$$

where R and S are, respectively, a 2- and 3-place relation symbol, f is a 2-place function symbol, c is a constant and x, y, z, v are variables. We work backwards as in Example 2.2.4. The last symbol is the variable v. Moving to the left we search for the rightmost symbol which is either a relation symbol or the symbol =. Here we find the ternary relation symbol S and the "subformula" $\langle S, x, y, v \rangle$ which we may write as $S(x, y, v)$. (Had we found the symbol =, then we would know that the part to the right of that symbol must be of the form $t * s$ for terms t, s; and we can recover t and s by Exercise 54 to find the "subformula" $\langle = \rangle * t * s$, which we may write as $(t = s)$.) Now note that formulas are constructed from the atomic ones by unary constructors (prefixing by $\forall x$, $\exists x$ or $\neg$) and binary constructors (combining φ, ψ to get $\langle \wedge \rangle * \varphi * \psi$ or similar, using $\vee$ or $\rightarrow$). Having obtained the subformula $S(x, y, v)$ we look at the string for the maximal number of unary constructors applied to it. In this example we find one: $\exists, v$ preceding our subformula. So we have a subformula $\langle \exists, v, S, x, y, v \rangle$ which we may write as $\exists v S(x, y, v)$. Next we move left to find the rightmost binary constructor. Here we find $\wedge$ and we know that the string between this $\wedge$ and our subformula must be a formula: it is $\langle \neg, =, f, y, z, c \rangle$, which we may write as $\neg(f(y, z) = c)$ and we have a subformula which may be written as $(\neg(f(y, z) = c) \wedge \exists v S(x, y, v))$. Continuing in this way we obtain in the end the formula

$$\exists x (R(x, y) \rightarrow \forall z (\neg(f(y, z) = c) \wedge \exists v S(x, y, v))).$$

We introduce therefore the following informal notation, which we will use throughout in this book.

Definition 2.2.9 (Notation of formulas)   Atomic formulas are written as $(t = s)$, $R(t_1, \ldots, t_n)$, $\bot$, R (for R a 0-ary relation symbol). Formulas constructed by clause (iv) of Definition 2.2.7 are written as $(\varphi \wedge \psi)$, $(\varphi \vee \psi)$, $(\varphi \rightarrow \psi)$ and $\neg \varphi$. Formulas constructed by clause (v) of Definition 2.2.7 are written as $\forall x \varphi$ and $\exists x \varphi$.

Exercise 57 Prove the following two statements simultaneously, by induction on strings:

(a) Whenever a string is a concatenation of two formulas, we can recover these two formulas uniquely from the string.
(b) Whenever a string is a formula, we can retrieve its informal notation uniquely from the string.

Examples/Remarks

(a) Note that, in contrast to the formal definition of formulas, in the informal notation we need brackets: $\varphi \vee \psi \rightarrow \chi$ might come from $\langle \vee \rangle * \varphi * \langle \rightarrow \rangle * \psi * \chi$, giving $\varphi \vee (\psi \to \chi)$; or from $\langle \to, \vee \rangle * \varphi * \psi * \chi$ which gives $(\varphi \vee \psi) \to \chi$. However, outermost brackets are usually omitted.

(b) Suppose the language L has one constant c, one binary function symbol f and one ternary relation symbol R. Then the expressions

$$\forall x \forall y R(c, x, f(y, c))$$
$$\forall x (x = f(x, x) \to \exists y R(x, c, y))$$
$$R(f(x, f(c, f(y, c))), c, y) \wedge (x = y \vee \neg R(c, c, x))$$

are formulas of L (note how we use the brackets!), but the expression

$$\forall R \neg R(x, x, c)$$

is not (this might be called a "second-order formula"; quantifying over relations).

There is also a *recursion principle* for defining functions on the set of formulas:

Proposition 2.2.10 (Recursion on formulas) *Let V be the set of variables, A the set of atomic formulas and F the set of all formulas. Suppose X is a set and we are given functions:*

$$f_a : A \to X$$
$$f_\wedge, f_\vee, f_\to : X \times X \to X$$
$$f_\neg : X \to X$$
$$f_\forall, f_\exists : V \times X \to X$$

Then there is a unique function $f : F \to X$ satisfying the following conditions:

$$f(\varphi) = f_a(\varphi) \text{ if } \varphi \text{ is an atomic formula.}$$
$$f(\varphi) = f_\wedge(f(\psi), f(\chi)) \text{ if } \varphi = \psi \wedge \chi.$$
$$f(\varphi) = f_\vee(f(\psi), f(\chi)) \text{ if } \varphi = \psi \vee \chi.$$
$$f(\varphi) = f_\to(f(\psi), f(\chi)) \text{ if } \varphi = \psi \to \chi.$$
$$f(\varphi) = f_\neg(f(\psi)) \text{ if } \varphi = \neg\psi.$$
$$f(\varphi) = f_\forall(x, f(\psi)) \text{ if } \varphi = \forall x \psi.$$
$$f(\varphi) = f_\exists(x, f(\psi)) \text{ if } \varphi = \exists x \psi.$$

Exercise 58 Prove Proposition 2.2.10. [Hint: argue in a way similar to the proof of Proposition 1.4.6.]

An important example of recursion on formulas is the definition of *substitution of a term for a variable* in a formula.

Definition 2.2.11 Let φ be a formula, x a variable and s a term. The *substitution of s for x in φ*, which we will temporarily denote by $\varphi[\![\,s/x\,]\!]$ in order to distinguish it from the substitution in Definition 2.2.6, is defined by recursion on φ as follows:

$$\varphi[\![\,s/x\,]\!] = \varphi[s/x] \text{ if } \varphi \text{ is an atomic formula; the}$$
$$\text{notation } \varphi[s/x] \text{ is from Definition 2.2.6.}$$
$$(\psi \circ \chi)[\![\,s/x\,]\!] = (\psi[\![\,s/x\,]\!] \circ \chi[\![\,s/x\,]\!]) \text{ for } \circ \in \{\wedge, \vee, \rightarrow\}.$$
$$(\neg\psi)[\![\,s/x\,]\!] = \neg\psi[\![\,s/x\,]\!].$$
$$(\forall y\psi)[\![\,s/x\,]\!] = \forall y\psi[\![\,s/x\,]\!] \text{ if } y \text{ is a variable distinct from } x.$$
$$(\forall x\psi)[\![\,s/x\,]\!] = \forall x\psi.$$
$$(\exists y\psi)[\![\,s/x\,]\!] = \exists y\psi[\![\,s/x\,]\!] \text{ if } y \text{ is a variable distinct from } x.$$
$$(\exists x\psi)[\![\,s/x\,]\!] = \exists x\psi.$$

From now on we shall write $\varphi[s/x]$ instead of $\varphi[\![\,s/x\,]\!]$.

Remark 2.2.12 Our treatment of terms and formulas above may seem a bit bureaucratic. Standard logic textbooks (e.g. [62]) deal with this very quickly. They say for example: *terms are strings of symbols. Every variable x and every constant c is a term. If f is an n-place function symbol and $t_1, \ldots, t_n$ are terms then $f(t_1, \ldots, t_n)$ is a term.* However, this does not make immediate mathematical sense. If the t_i are strings of symbols, then what *is* "$f(t_1, \ldots, t_n)$" and in what way is it a string of symbols (rather than, as it would appear, a string of symbols and strings (of symbols, and...))? As we have seen, the usual casual definition can be justified, and the more intuitive notation can be uniquely recovered from our formal one.

2.2.1   Free and Bound Variables

In the formula $\forall x(R(x, y) \rightarrow \exists z P(x, z))$ the variables x and z are *bound* by the quantifiers $\forall x$, $\exists z$ respectively; the variable y is *free*. The x in "$\forall x$" is also considered to be bound.

The intuition is that the formula above states a property of y but not of x or z; it should mean the same thing as the formula

$$\forall u(R(u, y) \rightarrow \exists v P(u, v)).$$

In the next section, where we shall give *meaning* to formulas, this will be made precise. The situation is similar to the use of variables in expressions such as $\int_0^x f(t)dt$: this expression is a function of x, not of t. And, of course, $\int_0^x f(t)dt = \int_0^x f(u)du$.

Now it may happen that the same variable occurs in more than one place, as in the formula $\forall y(R(x, y) \rightarrow \forall x R(x, x))$. Here the first *occurrence* of the variable x (in $\forall y R(x, y)$) is free whereas the two other occurrences (in $\forall x R(x, x)$) are bound. Let us make a definition.

Definition 2.2.13

  (i) Let φ be an L-formula. An *occurrence* of the variable x in φ is a natural number i such that the i-th element of the string φ is x.

 (ii) Let i be an occurrence of the variable x in formula φ, and let u be a variable which does not occur in φ. The occurrence i of x in φ is *bound* if the i-th element of the string $\varphi[u/x]$ is x (note that the strings φ and $\varphi[u/x]$ have the same length).

(iii) If the occurrence i of x in φ is not bound, it is called *free*.

(iv) A formula with no free (occurrences of) variables is called *closed*, or a *sentence*.

A closed formula should be thought of as an *assertion*.

Usually one talks about "free (and bound) variables" instead of "free and bound occurrences of variables". In practice we never consider expressions such as

$$\forall x \forall y \forall x\, R(x, y)$$

or

$$\forall y (R(x, y) \rightarrow \forall x\, R(x, x)).$$

In the first case one might be forgiven for thinking that the variable x is bound *twice*; and in the second case one has the feature that the variable x occurs both bound and free. When working with formulas we shall always stick to the following

CONVENTION ON VARIABLES *In formulas, a variable will always be either bound or free but not both; and every occurrence of it is bound at most once.*

This convention is not meant to exclude formulas like $\forall x\, P(x) \vee \neg \forall x\, P(x)$; certainly one can argue that the "same" variable (namely, x) is "bound twice"; but in fact every *occurrence* of the variable is only bound once. However, in the case of $\forall x (P(x) \vee \neg \forall x\, P(x))$ we shall rather use the equivalent form (see Section 2.3 for a precise definition of "equivalence of formulas") $\forall x (P(x) \vee \neg \forall y\, P(y))$.

    In the case of the two expressions above, we would replace $\forall x \forall y \forall x\, R(x, y)$ by $\forall x \forall y \forall u\, R(u, y)$ and $\forall y (R(x, y) \rightarrow \forall x\, R(x, x))$ by $\forall y (R(x, y) \rightarrow \forall u\, R(u, u))$.

2.2.2 Legitimate Substitutions

Substitution presents us with some subtleties. Suppose φ is the formula $\forall x\, R(x, y)$. If t is the term $f(u, v)$, then $\varphi[t/x]$ is just φ, since x is bound in φ; on the other hand, $\varphi[t/y]$ is $\forall x\, R(x, f(u, v))$. This is totally unproblematic. Intuitively, the formula φ states a property of y. The substitution $\varphi[t/y]$ should state the same property for t.

    But now suppose t is the term $f(x, y)$. The substitution $\varphi[t/y]$ presents us with a problem; if we carry out the replacement of y by t we get the formula $\forall x\, R(x, f(x, y))$, which intuitively does not "mean" that the property expressed by

φ holds for the element denoted by t! Therefore, we say that the substitution is not *legitimate* in this case.

Definition 2.2.14 A substitution $\varphi[t/x]$ is *legitimate* if no variable in the term t becomes bound in $\varphi[t/x]$.

The notion of legitimate substitution plays a role when we consider formal proofs in Chapter 3. In practice we shall consider the formula $\varphi = \forall x R(x, y)$ from the example above as the "same" formula as $\forall u R(u, y)$ (see Exercise 60 below), and now the substitution $\varphi[f(x, y)/y]$ makes sense: we get $\forall u R(u, f(x, y))$.

If the term t is closed (in particular, if t is a constant), the substitution $\varphi[t/x]$ is always legitimate, as is easy to see.

2.2.3   First-Order Logic and Other Kinds of Logic

In this book, we shall limit ourselves to the study of "first-order logic", which is the study of the formal languages and formulas as we have described here, and their relation to structures, as we will see in the next section.

This logic has good mathematical properties, but it also has severe limitations. Our variables denote, as we shall see, elements of structures. So we can only say things about *all elements* of a structure, not about all subsets, or about sequences of elements. For example, consider the language of orders: we have a 2-place relation symbol $\leq$ for "less than or equal to". We can express that $\leq$ really is a partial order:

$$(\forall x(x \leq x) \wedge$$
$$\forall x \forall y \forall z((x \leq y \wedge y \leq z) \rightarrow x \leq z)) \wedge$$
$$\forall x \forall y((x \leq y \wedge y \leq x) \rightarrow x = y)$$

and that $\leq$ is a linear order:

$$\forall x \forall y(x \leq y \vee y \leq x)$$

but we *cannot* express that $\leq$ is a well-order, since for that we have to say something about *all subsets* (we shall return to this example in Exercise 81).

It is possible to consider logics where such statements can be formed: these are called "higher-order" logics. There are also logics in which it is possible to form the conjunction, or disjunction, of an infinite set of formulas (so formulas will be infinite objects in such a logic). For a bit more on these logics, see Section A.5 in the Appendix.

2.3 Structures for First-Order Logic

In this section we consider a fixed but arbitrary first-order language L and discuss what it means to have a *structure for L*.

Definition 2.3.1 An *L-structure M* consists of a nonempty set, also denoted by M, together with the following data:

(i) for each constant c of L, an element c^M of M;
(ii) for each n-place function symbol f of L, a function

$$f^M : M^n \to M$$

(iii) for each n-place relation symbol R of L, a subset

$$R^M \subseteq M^n.$$

We call the element c^M the *interpretation* of c in M and similarly f^M and R^M are called the interpretations of f and R, respectively.

Given an L-structure M, we consider the language L_M (the *language of the structure M*): L_M is L together with, for each element m of M, an additional constant (also denoted m). Here it is assumed that $C_L \cap M = \emptyset$. If we stipulate that the interpretation in M of each new constant m is the element m, then M is also an L_M-structure.

Definition 2.3.2 (Interpretation of terms) For each closed term t of the extended language L_M we define its interpretation t^M as an element of M, by induction on t, as follows. If t is a constant, then its interpretation is already defined since M is an L_M-structure. If t is of the form $f(t_1, \ldots, t_n)$, then $t_1, \ldots, t_n$ are also closed terms of L_M, so by the induction hypothesis their interpretations $t_1^M, \ldots, t_n^M$ have already been defined; we put

$$t^M = f^M(t_1^M, \ldots, t_n^M).$$

Next, we define for a closed formula φ of L_M what it means to say that "φ is *true in M*" (other ways of saying this are: φ *holds* in M, or M *satisfies* φ). We have the following notation for this notion: $M \models \varphi$. For its negation we write $M \not\models \varphi$.

Definition 2.3.3 (Interpretation of formulas) For a closed formula φ of L_M, the relation $M \models \varphi$ is defined by induction on φ:

– If φ is a closed atomic L_M-formula we have the following possibilities: it is equal to $\bot$, it is of the form $(t_1 = t_2)$, or it is of the form $R(t_1, \ldots, t_n)$ with $t_1, t_2, \ldots, t_n$ closed terms. Define:

$$M \models \bot \ never \ \text{holds},$$
$$M \models (t_1 = t_2) \ \text{if and only if} \ t_1^M = t_2^M,$$
$$M \models R(t_1, \ldots, t_n) \ \text{if and only if} \ (t_1^M, \ldots, t_n^M) \in R^M,$$

where the t_i^M are the interpretations of the terms according to Definition 2.3.2, and R^M the interpretation of R in the structure M.

– If φ is of the form $(\varphi_1 \wedge \varphi_2)$ define

$$M \models \varphi \text{ if and only if } M \models \varphi_1 \text{ and } M \models \varphi_2.$$

– If φ is of the form $(\varphi_1 \vee \varphi_2)$ define

$$M \models \varphi \text{ if and only if } M \models \varphi_1 \text{ or } M \models \varphi_2$$

(the "or" is to be read as *inclusive*: as either... or..., or both).

– If φ is of the form $(\varphi_1 \rightarrow \varphi_2)$ define

$$M \models \varphi \text{ if and only if } M \models \varphi_2 \text{ whenever } M \models \varphi_1.$$

– If φ is of the form $(\neg \psi)$ define

$$M \models \varphi \text{ if and only if } M \not\models \psi.$$

– If φ is of the form $\forall x \psi$ define

$$M \models \varphi \text{ if and only if } M \models \psi[m/x] \text{ for all } m \in M.$$

– If φ is of the form $\exists x \psi$ define

$$M \models \varphi \text{ if and only if } M \models \psi[m/x] \text{ for some } m \in M.$$

In the last two clauses, $\psi[m/x]$ results by substitution of the new constant m for x in ψ.

In a way, this truth Definition 2.3.3 simply translates the formulas of L_M (and hence, of L) into ordinary language. For example, if R is a binary (2-place) relation symbol of L and M is an L-structure, then $M \models \forall x \exists y R(x, y)$ if and only if for each $m \in M$ there is an $n \in M$ such that $(m, n) \in R^M$. In other words, R^M contains the graph of a function $M \rightarrow M$ (if we use AC).

Remark 2.3.4 In modern textbooks on model theory one seldom finds the requirement that the underlying set of an L-structure must be nonempty, and certainly there are good reasons to allow the empty set as an L-structure (provided L has no constants): consider the definition of a substructure generated by a subset just before Exercise 95, and the proof of Theorem 2.8.3, where the nonemptiness requirement is quite inelegant. Nevertheless, we stick to the old-fashioned convention that structures must be nonempty because it is harder to formulate a sound and complete proof system for possibly empty structures. Indeed, the empty structure satisfies $\forall x \varphi$, for any formula φ.

2.3.1 Validity and Equivalence of Formulas

We introduce the following abbreviation: $\varphi \leftrightarrow \psi$ abbreviates $(\varphi \to \psi) \wedge (\psi \to \varphi)$.

Definition 2.3.5 An L-formula φ is called *valid* if for every L-structure M and every substitution of constants from M for the free variables of φ we have $M \models \varphi$. Two L-formulas φ and ψ are called *(logically) equivalent* if the formula $\varphi \leftrightarrow \psi$ is valid.

For example, the closed formula $\exists x (x = x)$ is valid, since structures are required to be nonempty.

The next couple of exercises provide you with a number of useful equivalences between formulas.

Exercise 59 Show that the following formulas are valid:

$(\varphi \vee (\psi \vee \chi)) \leftrightarrow ((\varphi \vee \psi) \vee \chi)$

$\varphi \leftrightarrow \neg\neg\varphi$

$\neg\varphi \leftrightarrow (\varphi \to \bot)$

$(\varphi \to \psi) \leftrightarrow (\neg\varphi \vee \psi)$

$(\varphi \vee \psi) \leftrightarrow \neg(\neg\varphi \wedge \neg\psi)$

$(\varphi \wedge \psi) \leftrightarrow \neg(\neg\varphi \vee \neg\psi)$

$\exists x \varphi \leftrightarrow \neg\forall x \neg\varphi$

$\forall x \varphi \leftrightarrow \neg\exists x \neg\varphi$

The equivalences $\neg(\varphi \vee \psi) \leftrightarrow (\neg\varphi \wedge \neg\psi)$ and $\neg(\varphi \wedge \psi) \leftrightarrow (\neg\varphi \vee \neg\psi)$ are called *De Morgan's Laws*.

$(\varphi \wedge (\psi \vee \chi)) \leftrightarrow ((\varphi \wedge \psi) \vee (\varphi \wedge \chi))$

$(\varphi \vee (\psi \wedge \chi)) \leftrightarrow ((\varphi \vee \psi) \wedge (\varphi \vee \chi))$

$(\varphi \to (\psi \vee \chi)) \leftrightarrow ((\varphi \to \psi) \vee (\varphi \to \chi))$

$(\varphi \to (\psi \wedge \chi)) \leftrightarrow ((\varphi \to \psi) \wedge (\varphi \to \chi))$

$((\varphi \vee \psi) \to \chi) \leftrightarrow ((\varphi \to \chi) \wedge (\psi \to \chi))$

In the following, assume that x does not occur in φ:

$(\varphi \to \exists x \psi) \leftrightarrow \exists x (\varphi \to \psi)$

$(\exists x \psi \to \varphi) \leftrightarrow \forall x (\psi \to \varphi)$

$(\forall x \psi \to \varphi) \leftrightarrow \exists x (\psi \to \varphi)$

Check for yourself that a formula like $\exists x \varphi \leftrightarrow \neg\forall x \neg\varphi$ does *not* violate our Convention on Variables!

Exercise 60 Suppose φ is a formula and u a variable that does not occur in φ. Show that the formulas $\forall v \varphi$ and $\forall u (\varphi[u/v])$ are equivalent.

Exercise 61 Show by counterexamples that the following sentences are not valid:

$$\exists v(\phi(v) \to \psi) \to (\exists v \phi(v) \to \psi)$$
$$((\forall x \phi(x)) \to \psi) \to \forall x (\phi(x) \to \psi)$$

Exercise 62 Prove that every formula is equivalent to a formula which starts with a string of quantifiers, followed by a formula in which no quantifiers occur. Such a formula is said to be *in prenex normal form*.

Exercise 63 Disjuction of formulas is associative up to logical equivalence (this is the first item of Exercise 59) so if we form a formula by iterated disjunction, it does not matter how we bracket it; we may write $\psi_1 \vee \cdots \vee \psi_n$. A similar remark goes for conjunction.

(a) Let φ be a formula in which no quantifiers occur. Show that φ is logically equivalent to a formula of the form:

$$\psi_1 \vee \cdots \vee \psi_k$$

where each ψ_i is a conjunction of atomic formulas and negations of atomic formulas. This form is called a *disjunctive normal form* for φ.

(b) Let φ be as in (a); show that φ is also equivalent to a formula of the form

$$\psi_1 \wedge \cdots \wedge \psi_k$$

where each ψ_i is a disjunction of atomic formulas and negations of atomic formulas. This form is called a *conjunctive normal form* for φ.

Exercise 64 Let L be a language with just one binary function symbol f. For each item in the list of L-structures given below, give an L-sentence which is true in that structure but false in the other three structures.

(a) $\mathbb{R} - \{0\}$ with the usual multiplication (as interpretation of f).
(b) $\{x \in \mathbb{R} \mid |x| > 1\}$ with the usual multiplication.
(c) $\{x \in \mathbb{C} \mid |x| > 1\}$ with the usual multiplication.
(d) The set of 2×2-matrices with real coefficients whose determinant has absolute value > 1, with matrix multiplication.

In the following exercises you are asked to give L-sentences which "express" certain properties of structures. This means: give an L-sentence ϕ such that for every L-structure M it holds that $M \models \phi$ if and only if the structure M has the given property.

Example 2.3.6 Suppose L has a 2-place function symbol m. An L-structure M is a nonempty set with a function $m^M : M \times M \to M$. If ϕ is the L-sentence

$$\forall x \forall y \forall z (m(x, m(y, z)) = m(m(x, y), z))$$

then $M \models \phi$ precisely when the operation m^M is associative. So we say that the sentence ϕ expresses the associativity of m.

Exercise 65 Let L be the empty language, so an L-structure is "just" a nonempty set M. Express by means of an L-sentence that M has exactly 4 elements.

Exercise 66 Let L be a language with one binary relation symbol R. Give L-sentences which express:

(a) R is an equivalence relation.
(b) There are exactly 2 equivalence classes.

Exercise 67 Let L be a language with just one 1-place function symbol F. Give an L-sentence ϕ which expresses that F is a bijective function.

Exercise 68 Let L be the language with just the 2-place function symbol written as a dot: $\cdot$. We consider the L-structures $\mathbb{Z}$ and $\mathbb{Q}$ where the dot $\cdot$ is interpreted as ordinary multiplication.

(a) "Define" the numbers 0 and 1. That is, give L-formulas $\varphi_0(x)$ and $\varphi_1(x)$ with one free variable x such that, in both $\mathbb{Q}$ and $\mathbb{Z}$, $\varphi_i(a)$ is true exactly when $a = i$ ($i = 0, 1$).
(b) Give an L-sentence which is true in $\mathbb{Z}$ but not in $\mathbb{Q}$.

Exercise 69 Let L be the language $\{f, g\}$ where f is a 2-place function symbol and g a 1-place function symbol. Consider the L-structure M, with underlying set $\mathbb{R}$, f^M is multiplication on $\mathbb{R}$, and g^M is the sine function. Give an L-formula $\phi(x)$ with one free variable x such that for all $a \in \mathbb{R}$ the following holds:

$$M \models \phi(a) \quad \Leftrightarrow \quad \text{there is an } n \in \mathbb{N} \text{ such that } a = (2n + \frac{1}{2})\pi.$$

Exercise 70 Let $L = \{\leq\}$ be the language of posets; here $\leq$ is a binary relation symbol (and we naturally write $x \leq y$ instead of $\leq (x, y)$). So a poset is nothing but an L-structure which satisfies the following L-sentences:

$$\forall x(x \leq x)$$
$$\forall x \forall y \forall z((x \leq y \wedge y \leq z) \rightarrow x \leq z)$$
$$\forall x \forall y((x \leq y \wedge y \leq x) \rightarrow x = y)$$

Suppose M is a well-order, seen as an L-structure. Give an L-formula $\phi(x)$ in one free variable such that for every $a \in M$ the following holds:

$$M \models \phi(a) \quad \Leftrightarrow \quad a \text{ is a limit element.}$$

2.4   Examples of Languages and Structures

2.4.1   Graphs

A directed graph is a structure with vertices (points) and edges (arrows) between them, such as:

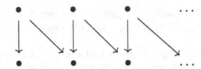

The language L_{graph} of directed graphs has two 1-place relation symbols, E and V (for "edge" and "vertex"), and two 2-place relation symbols S and T (for "source" and "target"; $S(x, y)$ will mean "the vertex x is the source of the edge y").

An L_{graph}-structure is a nonempty set G together with two subsets E^G, V^G of G, and two subsets S^G, T^G of G^2. G is a directed graph precisely when G satisfies the following *axioms* for directed graphs:

$\forall x (E(x) \vee V(x))$ $\forall x \neg (E(x) \wedge V(x))$

$\forall x \forall y (S(x, y) \rightarrow (V(x) \wedge E(y)))$ $\forall x \forall y (T(x, y) \rightarrow (V(x) \wedge E(y)))$

$\forall x \forall y \forall z ((S(x, z) \wedge S(y, z)) \rightarrow x = y)$ $\forall x \forall y \forall z ((T(x, z) \wedge T(y, z)) \rightarrow x = y)$

$\forall z (E(z) \rightarrow \exists x \exists y (S(x, z) \wedge T(y, z)))$

2.4.2   Local Rings

The language L_{rings} of rings has constants 0 and 1 and two 2-place function symbols for multiplication and addition, denoted $\cdot$ and $+$. There are no relation symbols.

A *commutative ring with* 1 is an L_{rings}-structure which satisfies the axioms for commutative rings with 1:

$\forall x (x + 0 = x)$ $\forall x (x \cdot 1 = x)$

$\forall x y (x + y = y + x)$ $\forall x y (x \cdot y = y \cdot x)$

$\forall x y z (x + (y + z) = (x + y) + z)$ $\forall x y z (x \cdot (y \cdot z) = (x \cdot y) \cdot z)$

$\forall x \exists y (x + y = 0)$ $\forall x y z (x \cdot (y + z) = x \cdot y + x \cdot z)$

$\qquad\qquad\qquad\qquad$ $\neg (0 = 1)$

We have started to abbreviate a string of quantifiers of the same kind: instead of $\forall x \forall y$ we write $\forall x y$.

A *local ring* is a commutative ring with 1 which has exactly one maximal ideal. This is a condition that involves quantifying over subsets (ideals) of the ring, and cannot be formulated in first-order logic. However, one can show that a commutative ring with 1 is local precisely when for each pair of elements x, y the statement that $x + y$ is a unit implies that either x or y is a unit. That is, a commutative ring R with 1 is local if and only if

$$R \models \forall xy (\exists z (z \cdot (x + y) = 1) \rightarrow (\exists v(v \cdot x = 1) \vee \exists w(w \cdot y = 1))).$$

Exercise 71 Let L be L_{rings} together with an extra unary relation symbol I. Give L-formulas which express that the subset defined by I is:

(a) an ideal;
(b) a prime ideal;
(c) a maximal ideal.

2.4.3 Vector Spaces

Fix a field k. We can write down a language L_k of first-order logic, and axioms in this language, such that the L_k-structures which satisfy the axioms are precisely the k-vector spaces.

The language L_k has a constant 0 and a binary function symbol $+$ to describe the abelian group structure. Furthermore, it has a 1-place function symbol f_m for every element m of k, to describe scalar multiplication. Apart from the axioms for an abelian group (which are the left side of the axioms for rings given in Section 2.4.2), there are the axioms:

$$f_m(0) = 0 \qquad\qquad \forall xy(f_m(x + y) = f_m(x) + f_m(y))$$
$$\forall x(f_1(x) = x) \qquad\qquad \forall x(f_m(f_{m'}(x)) = f_{mm'}(x))$$
$$\forall x(f_{m+m'}(x) = f_m(x) + f_{m'}(x))$$

In the second line of these axioms, 1 is the unit of the field k, and mm' refers to multiplication in k. In the third line, $m + m'$ refers to addition in k. Note that, if the field k is infinite, then there are infinitely many axioms to satisfy!

Exercise 72 The language L_k and the axioms for vector spaces given above are not very satisfactory in the sense that there are many important things about vectors that cannot be expressed by L_k-formulas; for example, that x and y are linearly independent vectors (if k is infinite).

Devise yourself a different language and different axioms which do allow you to express that two vectors are linearly independent over the field of scalars. Mimicking the example of graphs, have two 1-place relation symbols S and V (for "scalar" and "vector" respectively). How do you express addition of vectors and scalar multiplication?

2.4.4   Basic Plane Geometry

The language L_{geom} of basic plane geometry has two 1-place relation symbols P and L for "point" and "line", and a 2-place relation symbol I for "point x lies on line y". The axioms are:

$$\forall x(P(x) \vee L(x))$$
$$\forall x \neg(P(x) \wedge L(x))$$
$$\forall xy(I(x, y) \rightarrow (P(x) \wedge L(y)))$$
$$\forall xx'(P(x) \wedge P(x') \rightarrow \exists y(I(x, y) \wedge I(x', y)))$$
$$\forall xx'yy'((I(x, y) \wedge I(x', y) \wedge I(x, y') \wedge I(x', y')) \rightarrow (x = x' \vee y = y'))$$

Convince yourself that these axioms mean: everything is either a point or a line (and not both), for every two points there is a line they lie on, and two distinct lines can have at most one point in common.

Exercise 73 A famous extra axiom says that for every line l and point x not on l, there is a unique line m through x which does not intersect l. Show how to express this axiom in L_{geom}.

2.5   The Compactness Theorem

Before we can state the main theorem of this section we discuss some abstract general notions concerning first-order languages and structures.

Let L be a language. A *theory* in L (or L-theory) is simply a set of L-sentences (closed formulas). This may be a set of axioms for a meaningful mathematical theory, such as the axioms for local rings.

If T is an L-theory, an L-structure M is called a *model* of T if every sentence in T is true in M; in other words, if

$$M \models \varphi$$

for every $\varphi \in T$. We shall also write $M \models T$ in this case. So, a local ring is the same thing as a model of the theory of local rings, etcetera.

For every theory T there will be sentences which are true in every model of T: the consequences of the theory. We write $T \models \varphi$ to mean: φ holds in every model of T.

A theory T need not have models; T is said to be *consistent* if T has a model. The antonym is *inconsistent*.

Exercise 74 If T is inconsistent, $T \models \varphi$ holds for every L-sentence φ. Show also that $T \models \varphi$ if and only if $T \cup \{\neg\varphi\}$ is inconsistent.

Clearly, every model of T is also a model of every subtheory $T' \subseteq T$; so if T is consistent, so is T'. The following important theorem says that in order to check whether a theory T is consistent, one only needs to look at its *finite* subtheories:

Theorem 2.5.1 (Compactness Theorem; Gödel 1929) *Let T be a theory in a language L. If every finite $T' \subseteq T$ is consistent, then so is T.*

The easiest proof of the Compactness Theorem uses the Completeness Theorem (Theorem 3.2.6), which is proved in Chapter 3. However, it is not strictly necessary to go this way. If you would like to see a proof right away, see Section 2.5.1.

Exercise 75 Use the Compactness Theorem to show: if $T \models \varphi$ then there is a finite subtheory $T' \subseteq T$ such that $T' \models \varphi$.

The Compactness Theorem can be used to explore the boundaries of what can be expressed using first-order logic. Here are a few examples.

Example 2.5.2 Consider the empty language L: no constants, function symbols or relation symbols. An L-structure is nothing but a nonempty set. Still, there are meaningful L-sentences; for example the sentence

$$\forall xyz(x = y \lor x = z \lor y = z)$$

will be true in a set S if and only if S has at most two elements. Likewise, there is for any natural number $n \geq 1$ a sentence ϕ_n such that ϕ_n is true in S if and only if S has at most n elements.

Exercise 76 Give a construction for the sentences ϕ_n.

Consequently, if T is the theory $\{\neg \phi_n \mid n \geq 1\}$, then S is a model of T if and only if S is infinite.

In contrast, there is no theory T such that S is a model of T if and only if S is *finite*. This can be proved with the help of the Compactness Theorem. For, suppose that such a theory T exists. Consider then the theory

$$T' = T \cup \{\neg \phi_n \mid n \geq 1\}.$$

Assume that M is a model of T'. Then M is also a model of T, so M is finite by our assumption on T. Yet, by construction of T', M has at least $n + 1$ elements, for each natural number n! Clearly, this is impossible, so T' has no models.

But now by the Compactness Theorem, there must be a finite subtheory $T'' \subseteq T'$ such that T'' has no models. Consider such T''. Then for some $k \in \mathbb{N}$ we must have that

$$T'' \subseteq T \cup \{\neg \phi_n \mid 1 \leq n \leq k\}.$$

But any finite set with at least $k + 1$ elements is a model of $T \cup \{\neg\phi_n \mid 1 \leq n \leq k\}$, hence of T''. We have obtained a contradiction, showing that the assumed theory T does not exist.

Exercise 77 Conclude from this reasoning that there cannot be a single sentence ϕ in the empty language such that ϕ is true in a set S precisely when S is infinite.

Example 2.5.3 The language L_{grp} of groups has one constant e and one 2-place function symbol $\cdot$. The theory T_{grp} of groups consists of the sentences:

$$\forall x (e \cdot x = x) \qquad\qquad \forall x (x \cdot e = x)$$
$$\forall x y z (x \cdot (y \cdot z) = (x \cdot y) \cdot z) \quad \forall x \exists y (x \cdot y = e \wedge y \cdot x = e)$$

A group is nothing but an L_{grp}-structure which is a model of T_{grp}. Given a group G, an element g is said to have finite order if for some $n > 0$, $g^n = \underbrace{g \cdot \ldots \cdot g}_{n}$ is the unit element of the group. The least such n is in this case called the order of g.

For each $n \geq 2$, there is a sentence ϕ_n of L_{grp} such that for any group G it holds that $G \models \phi_n$ if and only if G has no non-trivial elements whose order is a divisor of n:

$$\forall x (\underbrace{x \cdot \ldots \cdot x}_{n} = e \rightarrow x = e).$$

Therefore, in complete analogy to the case with sets as structures for the empty language (Example 2.5.2), there is a theory T, with $T_{\text{grp}} \subseteq T$, such that the models of T are precisely the groups which do not contain elements of finite order (such as the group $\mathbb{Z}$).

And again, in contrast there is *no* theory T such that its models are precisely the groups which *do* contain elements of finite order. This is proved, using the Compactness Theorem, in a way completely analogous to Example 2.5.2, and is therefore left as an exercise.

Exercise 78 Carry out the proof of the last statement in Example 2.5.3.

There are many variations on Example 2.5.3. We mention one in the following exercise.

Exercise 79 Consider the language L_{graph} of directed graphs (Section 2.4.1). In a directed graph, a *cycle* is a sequence $(e_1, \ldots, e_k)$ of edges, such that for each i with $1 \leq i < k$, the source of e_{i+1} is the target of e_i, and moreover the target of e_k is the source of e_1. The number k is the *length* of the cycle.

(a) Show that for each $n \geq 1$ there is an L_{graph}-sentence ϕ_n which is true in a graph G exactly when G has no cycles of length n.
(b) Show that there is no theory T in L_{graph} such that the models of T are precisely the graphs which contain cycles.

(c) Show that there is no finite theory T in the language L_{graph} such that the models of T are precisely the graphs which have no cycles.

Example 2.5.4 This example and Example 2.5.5 illustrate another use of the Compactness Theorem: it can be used to show the existence of new models of certain theories. Technically, this example is a little different from the first two in that it uses an extension of the language by a constant.

The theory PA of *Peano Arithmetic* describes the basic structure of the natural numbers. The language has two constants 0 and 1 and two binary function symbols + and ·, and is therefore the same as the language for rings. PA has the following axioms:

$$\forall x \neg (x + 1 = 0) \qquad\qquad \forall x y (x + 1 = y + 1 \to x = y)$$
$$\forall x (x + 0 = x) \qquad\qquad \forall x (x \cdot 0 = 0)$$
$$\forall x y (x + (y + 1) = (x + y) + 1) \quad \forall x y (x \cdot (y + 1) = x \cdot y + x)$$

and in addition, there are the so-called *induction axioms*. Suppose φ contains the free variables $x, y_1, \ldots y_n$ and does not contain the variable u; then the following is an axiom of PA:

$$\forall y_1 \cdots y_n ((\varphi[0/x] \wedge \forall x (\varphi \to \varphi[x + 1/x])) \to \forall u \varphi[u/x]).$$

The theory PA is consistent since the ordinary set $\mathbb{N}$ of natural numbers, with the usual interpretation of the symbols 0, 1, +, ·, is a model of PA.

However, there are other models of PA. This can be seen with the help of the Compactness Theorem: consider the language L, which is the language of PA together with one extra constant c. Let T be the L-theory which has all the axioms of PA, and moreover all the axioms:

$$\neg (c = 0)$$
$$\neg (c = 1)$$
$$\neg (c = 1 + 1)$$
$$\neg (c = (1 + 1) + 1)$$
$$\vdots$$

Suppose T' is a finite subtheory of T. Then T' contains only finitely many of these new axioms. Therefore, we can always make $\mathbb{N}$ into an L-structure which is a model of T', by picking a large enough natural number for the interpretation of the constant c.

Therefore, every finite subtheory T' of T is consistent; by the Compactness Theorem, T is consistent. So T has a model M. Then M is, in particular, a model of PA. One can show that in every model of PA, the interpretations of the closed terms

$$0, 1, 1 + 1, (1 + 1) + 1, ((1 + 1) + 1) + 1, \ldots$$

are all distinct, so there is an injective function from $\mathbb{N}$ into M. Moreover, in M there is the element c^M which must be distinct from the elements 0^M, 1^M, $(1+1)^M$, ..., because M is a model of T.

The element c^M is called a *nonstandard number* and M is a *nonstandard model*.

Exercise 80

(a) Prove that PA $\models \forall x(x = 0 \lor \exists y(x = y + 1))$.
(b) Let M be a nonstandard model of PA. Prove that M contains an infinite descending chain: there are elements $c_0, c_1, \ldots$ in M such that $c_0 > c_1 > \cdots$.

The theory of models of PA is very interesting from the point of view of Model Theory, and also from the point of view of Gödel's famous *Incompleteness Theorems*. See the Appendix (Section A.1) for more on this.

Example 2.5.5 Consider the following language: the language $L_\mathbb{R}$ which has a constant r for every real number r, an n-place function symbol f for every function $f : \mathbb{R}^n \to \mathbb{R}$, and an n-place relation symbol R for every subset $R \subseteq \mathbb{R}^n$.

Clearly, interpreting everything by itself, $\mathbb{R}$ is an $L_\mathbb{R}$-structure. Let $T_\mathbb{R}$ be the set of all $L_\mathbb{R}$-sentences ϕ such that $\mathbb{R} \models \phi$. Then $\mathbb{R}$ is a model of $T_\mathbb{R}$.

Now just as in the previous example, we form a new language L out of $L_\mathbb{R}$ by adding one extra constant c, and we let T be the union of $T_\mathbb{R}$ with the set of new axioms:

$$\{c > n \mid n \in \mathbb{N}\}.$$

And just as in the previous example, we see that every finite subtheory of T is consistent. Therefore by the Compactness Theorem, T has a model $\mathcal{R}$.

$\mathcal{R}$ is a model of $T_\mathbb{R}$, and there is an embedding of $\mathbb{R}$ into $\mathcal{R}$; but moreover, $\mathcal{R}$ contains the "infinite" element $c^\mathcal{R}$. $\mathcal{R}$ is a field, because the axioms for a field are true in $\mathbb{R}$ and hence form part of $T_\mathbb{R}$. Let $d \in \mathcal{R}$ be the multiplicative inverse of $c^\mathcal{R}$. Then in $\mathcal{R}$, d is greater than 0, yet it is smaller than $\frac{1}{n}$ for each n! An element d with these properties is called a *nonstandard element*. We say that $\mathcal{R}$ is a *model for nonstandard analysis*.

Using a model for nonstandard analysis allows one to define concepts of ordinary analysis without using the usual ε-δ definitions. For example, a function $f : \mathcal{R} \to \mathcal{R}$ is continuous at $x \in \mathcal{R}$ if and only if for each nonstandard element d, the element $|f(x + d) - f(x)|$ is at most nonstandard.

Moreover, a nonstandard element d is thought of as an "infinitesimal" element, and in a model of nonstandard analysis, the differential quotient $\frac{df}{dx}$ is a "real" quotient (instead of a limit): one says that the function f is differentiable at x if and only if for any two nonstandard elements d and d', the expressions $\frac{f(x+d)-f(x)}{d}$ and $\frac{f(x+d')-f(x)}{d'}$ differ by at most a nonstandard element.

Nonstandard Analysis originated in Logic and was first developed by Abraham Robinson (1918–1974) (see [64]). It has now developed into a subfield of Analysis; for a recent introduction, see e.g. [25].

Another variation on the theme of the Compactness Theorem concerns well-orders.

Exercise 81 Let L be the language with just a 2-place relation symbol $<$ for "less than".

(a) Give an L-sentence ϕ such that the models of ϕ are precisely the linear orders.
(b) Show that there is no L-theory T such that the models of T are precisely the well-ordered sets.
    [Hint: Suppose that such a theory T exists. Let L' be the language obtained from L by adding infinitely many new constants $c_1, c_2, \ldots$. Let T' be the L'-theory which contains T and a set of sentences saying that "$c_1 > c_2 > \cdots$ is an infinite descending chain" (recall Proposition 1.4.2). Use the Compactness Theorem to obtain a contradiction.]
(c) Use the technique of part (b) (and the hint there) to prove that for every infinite well-order M there is an L-structure M' such that the following hold:

   (i) M and M' satisfy the same L-sentences.
   (ii) M' is not a well-order.

Here are some miscellaneous exercises about the Compactness Theorem.

Exercise 82 For sets X, let us write '$|X|$ is divisible by 3' if either $|X|$ is finite and divisible by 3, or X is infinite. Prove that there is no sentence ϕ in the empty language which expresses this property. [Hint: suppose such a sentence ϕ existed. Consider $\neg\phi$.]

Exercise 83 Let L be an arbitrary language. A class $\mathcal{M}$ of L-structures is called *elementary* if there is an L-theory T such that $\mathcal{M}$ is precisely the class of all models of T.

Suppose that for such a class $\mathcal{M}$ we have that both $\mathcal{M}$ and its complement are elementary. Prove that there is an L-sentence ϕ such that $\mathcal{M}$ is precisely the class of all L-structures which satisfy ϕ.

Exercise 84 In this exercise we use the Compactness Theorem to prove that every set X admits a linear order (that is, there is a linear order on X). You should compare this with Exercise 28, where we proved the same result using Zorn's Lemma.

(a) First prove the statement for every *finite* X, by induction on $|X|$.
(b) Now let X be arbitrary. Let L be the language with one 2-place relation symbol $<$ and constants $\{c_x \mid x \in X\}$. The L-theory T has the axioms for a linear order:

$$\forall x \neg(x < x)$$
$$\forall x \forall y \forall z((x < y \wedge y < z) \rightarrow x < z)$$
$$\forall x \forall y (x < y \vee x = y \vee y < x)$$

and moreover the axiom $\neg(c_x = c_y)$ for every pair x, y of distinct elements of X. Prove, using the Compactness Theorem, that T is consistent.
(c) Let M be a model of T. Show that M induces a linear order on X.

Exercise 85 Use the Compactness Theorem in order to show that for every partially ordered set $(X, \leq)$ there is a linear order $\preceq$ on X which extends $\leq$, that is: for all $x, y \in X$ we have $x \leq y \rightarrow x \preceq y$.

Exercise 86 Let L be the language of rings and ϕ an L-sentence. Suppose that for every natural number n there is a prime number $p > n$ and a field F of characteristic p such that $F \models \phi$. Show that there is a field K of characteristic 0 such that $K \models \phi$.

Exercise 87 (De Bruijn–Erdős) The result to be proved in this exercise was first published in [15]; evidently, the authors (Nicolaas Govert de Bruijn 1918–2012; Paul Erdős 1913–1996) were unaware of the force of the Compactness Theorem at the time.

We consider *simple undirected graphs*: a simple undirected graph has edges just as the directed graphs of Section 2.4.1, but now the edges have no direction. Moreover, a graph is simple if for any two vertices, there is at most one edge between them. In other words, a simple undirected graph is just a set with a symmetric binary relation. Let (X, R) be such a simple undirected graph, and k a positive integer. A k-*colouring* of (X, R) is a function f from X to the set $\{1, \ldots, k\}$ such that whenever $x, y \in X$ and $R(x, y)$ holds, $f(x) \neq f(y)$ (note that this implies that the relation R is irreflexive).

Prove (using the Compactness Theorem) the following statement: if every finite subgraph of (X, R) has a k-colouring, then (X, R) has a k-colouring.

2.5.1 Proof of the Compactness Theorem Using Ultrafilters

In this subsection we give a quite succinct proof (mainly in the form of exercises) of the Compactness Theorem.

Definition 2.5.6 Let I be a set. A *filter* on I is a collection $\mathcal{U}$ of subsets of I which satisfies the following conditions:

(i) $I \in \mathcal{U}$ and $\emptyset \notin \mathcal{U}$.
(ii) $\mathcal{U}$ is closed under binary intersection: if $A, B \in \mathcal{U}$ then $A \cap B \in \mathcal{U}$.
(iii) $\mathcal{U}$ is upwards closed: if $A \in \mathcal{U}$ and $A \subseteq B \subseteq I$, then $B \in \mathcal{U}$.

The set of filters on I is partially ordered by the subset relation: $\mathcal{U} \leq \mathcal{V}$ if $\mathcal{U} \subseteq \mathcal{V}$. A maximal element in the poset of filters on I is called an *ultrafilter* on I.

Exercise 88

(a) Let $\mathcal{U}$ be a filter on I and suppose that A is a subset of I such that $A \cap B \neq \emptyset$ for every $B \in \mathcal{U}$. Show that there is a filter $\mathcal{V}$ on I such that $A \in \mathcal{V}$ and $\mathcal{U} \leq \mathcal{V}$.
(b) Use Zorn's lemma in order to prove that every filter on I is contained in an ultrafilter on I.
(c) Let $\mathcal{U}$ be an ultrafilter on I. Show that for every subset A of I, either $A \in \mathcal{U}$ or $(I - A) \in \mathcal{U}$ (and not both).

(d) Let $\mathcal{U}$ be an ultrafilter on I. Show that for every pair A, B of subsets of I we have: if $A \cup B \in \mathcal{U}$ then $A \in \mathcal{U}$ or $B \in \mathcal{U}$.

Example 2.5.7 A filter $\mathcal{U}$ is *principal* if there is some $A \subseteq I$ such that $\mathcal{U}$ is precisely the set of subsets of I which contain A. Let us say that $\mathcal{U}$ is *generated* by A in this case.

Exercise 89

(a) Show that if $\mathcal{U}$ is generated by A then there is a bijection between the set of ultrafilters on I which contain $\mathcal{U}$ as subset, and the set of ultrafilters on A.
(b) Show that a principal ultrafilter is always generated by a singleton set (a set with one element).
(c) Show that if $\mathcal{U}$ is a non-principal ultrafilter, then $\mathcal{U}$ does not contain any finite set.

The main example of a non-principal filter on I (if I is infinite) is the set of cofinite subsets of I. This is called the *Fréchet filter* on I. By Exercise 89 we have that an ultrafilter on I is non-principal if and only if it extends the Fréchet filter on I.

Definition 2.5.8 Let $(A_i)_{i \in I}$ be a family of sets, and $\mathcal{U}$ an ultrafilter on I. The *ultraproduct* $\prod_{\mathcal{U}} A_i$ is the set of equivalence classes of the ordinary product $\prod_{i \in I} A_i$ by the following equivalence relation:

$$\alpha \sim_{\mathcal{U}} \beta \text{ if and only if } \{i \in I \mid \alpha_i = \beta_i\} \in \mathcal{U}.$$

We shall write $\mathcal{A}$ for $\prod_{\mathcal{U}} A_i$ and denote the equivalence class of $\alpha \in \mathcal{A}$ by $[\alpha]$. If, moreover, all A_i are L-structures, then we have an L-structure on $\mathcal{A}$ as follows.

For a constant c of L we let $c^{\mathcal{A}}$ be the equivalence class of the element $(c^{A_i})_{i \in I}$ of $\prod_{i \in I} A_i$.

For an n-place function symbol f of L and $[\alpha_1], \ldots, [\alpha_n] \in \mathcal{A}$ we let $f^{\mathcal{A}}([\alpha_1], \ldots, [\alpha_n])$ be the equivalence class of

$$(f^{A_i}((\alpha_1)_i, \ldots, (\alpha_n)_i))_{i \in I}.$$

(Check that this is well-defined on equivalence classes.)

For an n-place relation symbol R of L we let $([\alpha_1], \ldots, [\alpha_n]) \in R^{\mathcal{A}}$ precisely when the set $\{i \in I \mid ((\alpha_1)_i, \ldots, (\alpha_n)_i) \in R^{A_i}\}$ is an element of the ultrafilter $\mathcal{U}$. Again, here it should be checked that this is a valid definition.

Exercise 90 (Łoś's Theorem) (Jerzy Łoś, 1920–1998)

(a) Check that Definition 2.5.8 defines a correct L-structure on $\mathcal{A} = \prod_{\mathcal{U}} A_i$.
(b) Show that for every L-formula $\varphi(x_1, \ldots, x_n)$ and for every n-tuple $([\alpha_1], \ldots, [\alpha_n])$ of elements of $\mathcal{A}$ we have

$\mathcal{A} \models \varphi([\alpha_1], \ldots, [\alpha_n])$ if and only if the set

$$\{i \in I \mid A_i \models \varphi((\alpha_1)_i, \ldots, (\alpha_n)_i)\}$$

is an element of the ultrafilter $\mathcal{U}$.

Theorem 2.5.9 (Compactness Theorem) *Let T be an L-theory such that every finite subtheory of T has a model. Then T has a model.*

Proof. Let I be the set of all finite subsets of T. For an element $i = \{\phi_1, \ldots, \phi_n\}$ of I, pick a model A_i of $\{\phi_1, \ldots, \phi_n\}$. For every L-sentence ϕ let $\hat{\phi}$ be the set $\{i \in I \mid \phi \in i\}$. Note that for an n-tuple $(\phi_1, \ldots, \phi_n)$ of L-sentences we have

$$\{\phi_1, \ldots, \phi_n\} \in \hat{\phi}_1 \cap \cdots \cap \hat{\phi}_n$$

so this intersection is nonempty. Therefore the set of those subsets of I which contain some $\hat{\phi}_1 \cap \cdots \cap \hat{\phi}_n$ as a subset, is a filter on I. Let U be an ultrafilter which extends this filter and let $\mathcal{A} = \prod_{\mathcal{U}} A_i$ as an L-structure according to Definition 2.5.8. We see that for every L-sentence ϕ the set $\{i \mid A_i \models \phi\}$ contains the set $\hat{\phi}$ as a subset, and is therefore an element of $\mathcal{U}$. It follows from Łoś's Theorem (Exercise 90) that $\mathcal{A} \models \phi$. Since ϕ was an arbitrary element of T, $\mathcal{A}$ is a model of T, as desired. ∎

2.6   Substructures and Elementary Substructures

Definition 2.6.1 (Isomorphism of L-structures)   Let M and N be L-structures. An *isomorphism* from M to N, or L-*isomorphism*, is a bijection $\beta : M \to N$ such that the following hold:

(i) $\beta(c^M) = c^N$ for every constant c of L.
(ii) $\beta(f^M(a_1, \ldots, a_n)) = f^N(\beta(a_1), \ldots, \beta(a_n))$ for every n-place function symbol f of L and every $a_1, \ldots, a_n \in M$.
(iii) $(a_1, \ldots, a_n) \in R^M \Leftrightarrow (\beta(a_1), \ldots, \beta(a_n)) \in R^N$ for every n-place relation symbol of L and $a_1, \ldots, a_n \in M$.

We say that M and N are *isomorphic* if there is an isomorphism $\beta : M \to N$.

It is easy to see that if $\beta : M \to N$ is an isomorphism then so is $\beta^{-1} : N \to M$, so the relation of being isomorphic is symmetric.

Exercise 91 Let $\varphi(y_1, \ldots, y_n)$ be an L-formula and $a_1, \ldots, a_n \in M$. Consider an isomorphism $\beta : M \to N$. Show that $M \models \varphi(a_1, \ldots, a_n)$ if and only if $N \models \varphi(\beta(a_1), \ldots, \beta(a_n))$. Conclude that isomorphic L-structures satisfy the same L-sentences.

Exercise 92 For a field k, let L_k be the language of k-vector spaces as in Section 2.4.3. Show that for k-vector spaces M and N, an L_k-isomorphism from M to N is the same thing as a bijective k-linear map. Show also that for two rings R and S, an L_{rings}-isomorphism from R to S is the same thing as a ring isomorphism. The same holds for graphs, groups, posets, etc.

Definition 2.6.2 Let M and N be structures for a language L. We say that N is a *substructure* of M, and write $N \subseteq M$, if N is a subset of M, and the following conditions are satisfied:

 (i) For every constant c of L, $c^N = c^M$.
 (ii) For every n-place function symbol f of L, $f^N : N^n \to N$ is the restriction of f^M to N^n (recall that this means that for all $a_1, \ldots, a_n \in N$, $f^M(a_1, \ldots, a_n)$ is an element of N, and equal to $f^N(a_1, \ldots, a_n)$).
 (iii) For every n-place relation symbol R of L, $R^N = R^M \cap N^n$ (this means that for $a_1, \ldots, a_n \in N$, $(a_1, \ldots, a_n) \in R^M$ if and only if $(a_1, \ldots, a_n) \in R^N$).

When we are considering models M, N of an L-theory T, we also say that N is a *submodel* of M if N is a substructure of M.

Exercise 93 Let $N \subseteq M$ be a substructure. Show that for every quantifier-free L-formula (that is, a formula without quantifiers) φ with variables $x_1, \ldots, x_n$ and for every n-tuple $m_1, \ldots, m_n$ of elements of N, we have

$$N \models \varphi(m_1, \ldots, m_n) \text{ if and only if } M \models \varphi(m_1, \ldots, m_n).$$

Exercise 94 Let N be an L-structure. The *diagram* of N, denoted $D(N)$, is the set of all quantifier-free L_N-sentences which are true in N.

(a) Suppose M is a model of $D(N)$. Show that M has a substructure which is isomorphic to N.
(b) Conversely, suppose that M is an L-structure such that N is isomorphic to a submodel of M. Show that M can be made into an L_N-structure which is a model of $D(N)$.

If M is an L-structure and $N \subseteq M$ is a nonempty subset which contains all the elements c^M and is closed under the functions f^M, then there is a unique way of making N into a substructure of M, by defining

$$R^N = R^M \cap N^n$$

for each n-place relation symbol R of L.

Now suppose $\{N_i \mid i \in I\}$ is a family of subsets of M such that each N_i contains all the constants c^M and is closed under the functions f^M. Then this also holds for the intersection $\bigcap_{i \in I} N_i$. Therefore, if M is an L-structure and S is an arbitrary nonempty subset of M, there is a *least* substructure of M which contains S as a subset; we shall call this the substructure *generated by* S. Note that the requirement that S be nonempty cannot be dropped in the case when L has no constants.

Exercise 95

(a) Show that the substructure generated by S can also be constructed as the union of a chain of subsets of M, as follows. Let S_0 be the union of S and the set $\{c^M \mid c$ a constant of $L\}$. Suppose $S \subseteq S_0 \subseteq S_1 \subseteq \cdots \subseteq S_k$ have been constructed; let S_{k+1} be the union of S_k and the set consisting of all elements of the form $f^M(a_1, \ldots, a_n)$ for an n-place function symbol of L and $a_1, \ldots, a_n \in S_k$.

(b) Write $|L|$ for the cardinality $|con(L) \cup fun(L) \cup rel(L)|$. Conclude from part (a) that the cardinality of the substructure generated by S is $\leq \max(|S|, |L|, \omega)$. In particular, if the language L is countable and S is countable, the substructure generated by S is countable too.

(c) More directly, the substructure generated by S is the set

$$\{t^M(a_1, \ldots, a_n) \mid t \text{ an } L\text{-term}, a_1, \ldots, a_n \in S\}.$$

Definition 2.6.3 A substructure $N \subseteq M$ is called an *elementary substructure*, written $N \preceq M$, if the equivalence of Exercise 93 holds for *all* L-formulas φ. Equivalently, if for every sentence φ of L_N we have:

$$N \models \varphi \text{ if and only if } M \models \varphi.$$

The notation $N \preceq M$ should not be confused with the same notation for embeddings between well-orders in Chapter 1.

The notion of "elementary substructure" means that, from the point of view of L, the elements of N have the same properties in N as in M. For example, consider $\mathbb{Q} \subseteq \mathbb{R}$ as a subring. Then this is *not* an elementary substructure, for 2 is a square in $\mathbb{R}$ but not in $\mathbb{Q}$. However, if we consider $\mathbb{Q}$ and $\mathbb{R}$ just as ordered structures (as structures for the language with just one binary relation symbol $<$), then $\mathbb{Q}$ is an elementary substructure of $\mathbb{R}$. We shall not prove this last fact here, but anticipating some definitions yet to come, we point out that the so-called *theory of dense linear orders without end-points* (see Definition 2.9.4 at the end of this chapter), of which both $\mathbb{Q}$ and $\mathbb{R}$ are models, has quantifier elimination (see the definition at the beginning of the next section). Hence the statement follows from Exercise 98 below.

Exercise 96 Suppose N is an L-structure. The *elementary diagram* of N, denoted $E(N)$, is the set of all L_N-sentences which are true in N. In analogy to Exercise 94, prove the following:

(a) Suppose M is a model of $E(N)$. Show that M has an elementary substructure which is isomorphic to N.

(b) Conversely, suppose that M is an L-structure such that N is isomorphic to an elementary submodel of M. Show that M can be made into an L_N-structure which is a model of $E(N)$.

In the literature the theory $E(N)$ is often denoted by $\text{Diag}_{el}(N)$.

Exercise 97 (Tarski–Vaught Test)   (Robert Lawson Vaught, 1926–2002) Suppose $N \subseteq M$ is an L-substructure. Show that $N \preceq M$ if and only if the following condition holds: for every L_N-sentence of the form $\exists x \varphi$ which is true in M, there exists an $a \in N$ such that $M \models \varphi[a/x]$. [Hint: use induction on L_N-sentences. Convince yourself that only the cases $\exists$, $\wedge$ and $\neg$ have to be considered.]

Note that, if $N \preceq M$ then in particular both structures satisfy the same L-sentences; hence for every L-theory T, the structure N is a model of T if and only if M is.

2.7 Quantifier Elimination

Definition 2.7.1 Let T be a theory in a language L. We say that T *admits elimination of quantifiers*, or *has quantifier elimination*, if for every L-formula φ with free variables $x_1, \ldots, x_n$ there is a *quantifier-free* L-formula ψ with at most the free variables $x_1, \ldots, x_n$, such that

$$T \models \forall x_1 \cdots x_n (\varphi \leftrightarrow \psi).$$

We also say that φ and ψ are T-*equivalent*.

In particular, if φ is a sentence, there will be a quantifier-free L-sentence ψ such that $T \models \varphi \leftrightarrow \psi$.

Exercise 98 Suppose the theory T admits elimination of quantifiers. Suppose M and N are models of T and $N \subseteq M$. Then N is an elementary substructure of M.

Exercise 99 Let L be a language, T an L-theory, M a model of T and A a substructure of M. Assume that the theory T admits quantifier elimination.

(a) Show that for every L_A-sentence ϕ there is an L_A-sentence ψ such that the following equivalence holds:

$$M \models \phi \iff A \models \psi.$$

(b) Now suppose that M_1 and M_2 are models of T, and A a substructure both of M_1 and M_2. Show that M_1 and M_2 satisfy the same L_A-sentences.

Applications of quantifier elimination often concern *completeness* of the theory T. We say that a theory T is *complete* if for every L-sentence φ, either $T \models \varphi$ or $T \models \neg\varphi$ holds. Clearly, if T admits quantifier elimination, then this condition only has to be checked for quantifier-free L-sentences φ.

Exercise 100 Show that T is complete if and only if any two models of T satisfy the same L-sentences.

The following lemma says that in order to check whether T has quantifier elimination, we may restrict ourselves to very simple formulas. Call a formula *simple* if it is of the form

$$\exists x(\psi_1 \wedge \cdots \wedge \psi_n \wedge \neg\chi_1 \wedge \cdots \wedge \neg\chi_m),$$

where $\psi_1, \ldots, \psi_n, \chi_1, \ldots, \chi_m$ are atomic formulas.

Lemma 2.7.2 *T admits elimination of quantifiers if and only if every simple formula is T-equivalent to a quantifier-free formula in at most the same free variables.*

Proof. Clearly, the given condition is necessary. In order to see that it is also sufficient, we argue by induction on φ to show that every φ is T-equivalent to a quantifier-free formula.

This is plainly true for atomic φ, and it is left to you to see that the set of formulas which are T-equivalent to a quantifier-free formula is closed under the operations $\wedge, \vee, \rightarrow$ and $\neg$.

For the quantifier case we use Exercise 63, which states that every quantifier-free formula is equivalent to a formula of the form

$$\psi_1 \vee \cdots \vee \psi_n,$$

where each ψ_i is a conjunction of atomic formulas and negated atomic formulas. Hence, if φ is T-equivalent to a quantifier-free formula, it is T-equivalent to one in this form, whence $\exists x\varphi$ is equivalent to $(\exists x\psi_1) \vee \cdots \vee (\exists x\psi_n)$, that is: a disjunction of simple formulas. Now the condition in the lemma tells us that each of these is T-equivalent to a quantifier-free formula, and therefore so is $\exists x\varphi$.

For the case $\forall x\varphi$ one simply uses that this is equivalent to $\neg\exists x\neg\varphi$. ∎

In this section, by way of example, we shall prove for one theory that it has quantifier elimination: the *theory of algebraically closed fields* T_{acf}. Recall that a field k is algebraically closed if every monic polynomial with coefficients in k has a root (a zero) in k. That this theory has quantifier elimination was proved by Alfred Tarski in 1948.

Let us use L for the language L_{rings}: the language of commutative rings with 1. The L-theory T_{acf} has, besides the axioms for commutative rings with 1, the axioms:

$$\forall x(\neg(x = 0) \rightarrow \exists y(x \cdot y = 1))$$
$$\forall y_1 \cdots y_n \exists x(x^n + y_1 \cdot x^{n-1} \cdots + y_{n-1} \cdot x + y_n = 0).$$

Here x^n abbreviates the term $\underbrace{x \cdots \cdots x}_{n \text{ times}}$.

The second line in the axioms for T_{acf} describes an axiom for each $n \geq 1$, so there are infinitely many axioms.

These axioms express that we have a field in which every monic polynomial has a root. In other words, an algebraically closed field.

Note that every term $t(x, y_1, \ldots, y_n)$ of L in variables $x, y_1, \ldots, y_n$ denotes a polynomial in the same variables with coefficients in $\mathbb{N}$, so with every atomic formula $t = s$ in these variables we can associate a polynomial P with coefficients in $\mathbb{Z}$ such that for every ring R and for all $a, b_1, \ldots, b_n \in R$,

$$R \models (t = s)(a, b_1, \ldots, b_n) \text{ iff } P(a, b_1, \ldots, b_n) = 0 \text{ in } R.$$

Furthermore, we notice that since every field is an integral domain, a conjunction $r_1 \neq 0 \wedge \cdots \wedge r_k \neq 0$ is equivalent to $r_1 \cdots r_k \neq 0$. So we can write every simple L-formula as

$$\exists x (P_1(x, y_1, \ldots, y_n) = 0 \wedge \cdots \wedge P_k(x, y_1, \ldots, y_n) = 0 \wedge Q(x, y_1, \ldots, y_n) \neq 0).$$

Definition 2.7.3 Let L be a language, Γ a set of L-formulas, M and N L-structures and $\vec{a} = a_1, \ldots, a_n$ and $\vec{b} = b_1, \ldots, b_n$ tuples of elements of M and N, respectively. Write $\vec{a} \equiv_\Gamma \vec{b}$ if for every formula $\phi(x_1, \ldots, x_n)$ from Γ we have:

$$M \models \phi(a_1, \ldots, a_n) \iff N \models \phi(b_1, \ldots, b_n).$$

We shall apply this for Γ the set of quantifier-free L-formulas and for Γ the set of simple L-formulas, in which case we write $\vec{a} \equiv_{qf} \vec{b}$, $\vec{a} \equiv_{simple} \vec{b}$, respectively.

Lemma 2.7.4 *Let L be an arbitrary language. Suppose that an L-theory T has the following property:*

Whenever M and N are models of T, and $\vec{a} = a_1, \ldots, a_n$, $\vec{b} = b_1, \ldots, b_n$ are tuples of elements of M and N, respectively, then $\vec{a} \equiv_{qf} \vec{b}$ implies $\vec{a} \equiv_{simple} \vec{b}$.

Then T has quantifier elimination.

Proof. Assume that T has the property in the statement of the Lemma. By Lemma 2.7.2 we have to show that every simple L-formula is T-equivalent to a quantifier-free formula in the same free variables. So, let $\exists v \phi(v, \vec{w})$ be a simple formula, with $\vec{w} = w_1, \ldots, w_n$ the free variables. Let $\vec{c} = c_1, \ldots, c_n$ be new constants; we write $L_{\vec{c}}$ for $L \cup \{c_1, \ldots, c_n\}$.

Let Γ be the set of all quantifier-free L-formulas $\psi(\vec{w})$ such that

$$T \models (\exists v \phi(v, \vec{c})) \rightarrow \psi(\vec{c})$$

and write $\Gamma(\vec{c})$ for $\{\psi(\vec{c}) \mid \psi(\vec{w}) \in \Gamma\}$.

Claim 1 $T \cup \Gamma(\vec{c}) \models \exists v \phi(v, \vec{c})$.

Let us first show that Claim 1 implies that the formula $\exists v \phi(v, \vec{w})$ is T-equivalent to a quantifier-free formula. We apply the Compactness Theorem and see that there must be finitely many $\gamma_1(\vec{c}), \ldots, \gamma_m(\vec{c}) \in \Gamma(\vec{c})$ such that

$$T \models \bigwedge_{i=1}^{m} \gamma_i(\vec{c}) \rightarrow \exists v \phi(v, \vec{c}).$$

This means, since the constants $\vec{c}$ do not appear in T, that

$$T \models \forall \vec{w} (\bigwedge_{i=1}^{m} \gamma_i(\vec{w}) \rightarrow \exists v \phi(v, \vec{w})).$$

Since all γ_i are elements of Γ, we see that the formula $\exists v \phi(v, \vec{w})$ is T-equivalent to the quantifier-free formula $\bigwedge_{i=1}^{m} \gamma_i(\vec{w})$, as desired.

To prove Claim 1, suppose for a contradiction that M is a model of $T \cup \Gamma(\vec{c})$ and $M \models \neg \exists v \phi(v, \vec{c})$. Let Δ be the set of all quantifier-free $L_{\vec{c}}$-sentences which are true in M.

Claim 2 The theory $T \cup \Delta \cup \{\exists v \phi(v, \vec{c})\}$ is consistent.

Proof of Claim 2: suppose that this theory is inconsistent, then by the Compactness Theorem there are finitely many elements $\delta_1(\vec{c}), \ldots, \delta_k(\vec{c})$ of Δ such that

$$T \cup \{\delta_1(\vec{c}), \ldots, \delta_k(\vec{c})\} \cup \{\exists v \phi(v, \vec{c})\}$$

is inconsistent. This means that

$$T \models \exists v \phi(v, \vec{c}) \rightarrow \neg \delta_1(\vec{c}) \vee \cdots \vee \neg \delta_k(\vec{c})$$

and therefore by definition of Γ, that the formula $\neg \delta_1(\vec{w}) \vee \cdots \vee \neg \delta_k(\vec{w})$ is an element of Γ.

Now M is, by assumption, a model of $\Gamma(\vec{c})$, so we have

$$M \models \neg \delta_1(\vec{c}) \vee \cdots \vee \neg \delta_k(\vec{c}).$$

On the other hand, the sentences $\delta_1(\vec{c}), \ldots, \delta_k(\vec{c})$ are elements of Δ and are therefore true in M by definition of Δ. Clearly, we have a contradiction now, which proves Claim 2.

Having proved Claim 2, we return to the proof of Claim 1. By Claim 2, let N be a model of $T \cup \Delta \cup \{\exists v \phi(v, \vec{c})\}$.

Let $\vec{a} = \vec{c}^M$ and $\vec{b} = \vec{c}^N$. We have now, for every quantifier-free L-formula $\psi(\vec{w})$:

$$M \models \psi(\vec{a}) \Leftrightarrow M \models \psi(\vec{c})$$
$$\Leftrightarrow \psi(\vec{c}) \in \Delta$$
$$\Leftrightarrow N \models \psi(\vec{c})$$
$$\Leftrightarrow N \models \psi(\vec{b}).$$

We conclude that $\vec{a} \equiv_{qf} \vec{b}$. However, $M \models \neg \exists v \phi(v, \vec{a})$ whereas $N \models \exists v \phi(v, \vec{b})$. Since $\exists v \phi(v, \vec{w})$ was assumed to be a simple formula, we see that $\vec{a} \not\equiv_{simple} \vec{b}$.

But M and N are models of T. So we see that T does *not* have the property in the statement of the Lemma. This contradiction proves Claim 1. ∎

In order to prove that the theory T_{acf} has quantifier elimination, we need one ingredient from algebra:

Fact. For any field k there is an algebraically closed field $\bar{k}$, the *algebraic closure of k*, such that $k \subset \bar{k}$ and moreover, whenever k is embedded in an algebraically closed field l, there is a (non-unique) extension of this embedding to an embedding of $\bar{k}$ into l. In that case, the image of $\bar{k}$ in l consists precisely of those elements which are zeros of polynomials with coefficients in k.

Theorem 2.7.5 (Tarski) *The theory T_{acf} has quantifier elimination.*

Proof. We wish to apply Lemma 2.7.4. Suppose k and k' are algebraically closed fields and $\vec{a} \in k$ and $\vec{b} \in k'$ are such that for every quantifier-free $L = L_{rings}$-formula $\psi(\vec{w})$ we have $k \models \psi(\vec{a})$ if and only if $k' \models \psi(\vec{b})$. Then the subring of k generated by $\vec{a}$ is isomorphic to the subring of k' generated by $\vec{b}$, so we may as well assume that $\vec{a} = \vec{b} \in k \cap k'$. Let $R \subset k \cap k'$ be the quotient field of the subring of $k \cap k'$ generated by $\vec{a}$.

Now let $\exists v \phi(v, \vec{w})$ be a simple L-formula, which we have seen may be taken to be of the form

$$\exists v (P_1(v, \vec{w}) = 0 \wedge \cdots \wedge P_k(v, \vec{w}) = 0 \wedge Q(v, \vec{w}) \neq 0),$$

where $P_1, \ldots, P_k, Q$ are polynomials with coefficients in $\mathbb{Z}$. We have to prove: if $k \models \exists v \phi(v, \vec{a})$ then $k' \models \exists v \phi(v, \vec{a})$.

If all the polynomials $P_i(v, \vec{a})$ are identically zero, then this reduces to the following statement: if $k \models \exists v Q(v, \vec{a}) \neq 0$ then $k' \models \exists v Q(v, \vec{a}) \neq 0$. But if $k \models \exists v Q(v, \vec{a}) \neq 0$, then the polynomial $Q(v, \vec{a})$ is not identically zero, and therefore has only finitely many zeros. On the other hand k', being algebraically closed, is infinite; hence $k' \models \exists v Q(v, \vec{a}) \neq 0$ as desired.

If not all polynomials P_i are identically zero, and $c \in k$ satisfies $k \models \phi(c, \vec{a})$, then c is algebraic over $\vec{a}$ and therefore an element of the algebraic closure of R. Since this algebraic closure embeds into k', we also have an element d of k' such that $k' \models \phi(d, \vec{a})$. We have verified the hypothesis of Lemma 2.7.4 and conclude that T_{acf} has quantifier elimination. ∎

Remark 2.7.6 In our quantifier elimination proof for T_{acf} we have made use of the Axiom of Choice: Lemma 2.7.4 relies on the Compactness Theorem, which needs Zorn's Lemma; and also the statement that every field has an algebraic closure cannot be proved without Choice. It is possible to prove quantifier elimination for T_{acf} without using Choice, as is done in the book [45]. However, Lemma 2.7.4 is also very useful for other quantifier elimination proofs.

Exercise 101 Let ϕ be the L_{rings}-sentence

$$\exists x (x^2 + 1 = 0 \wedge x + 1 \neq 0).$$

Give a quantifier-free L_{rings}-sentence ψ which is T_{acf}-equivalent to ϕ.

Exercise 102 Let k be an algebraically closed field, and $\phi(v)$ an L_{rings}-formula in one free variable v. Prove that the set $\{a \in k \mid k \models \phi(a)\}$ is either finite or cofinite.

In the book [49] you will find many more proofs of quantifier elimination for various theories.

Applications of Quantifier Elimination for T_{acf}

In this subsection we present a few mathematical applications of quantifier elimination for algebraically closed fields.

The theory T_{acf} is not complete, because it does not settle all quantifier-free sentences of L_{rings}: for example, the sentence $1 + 1 + 1 = 0$. However, once we specify the characteristic of the field, the theory becomes complete. Let ϕ_n be the sentence $\underbrace{1 + \cdots + 1}_{n \text{ times}} = 0$. Define the following theories:

$T_{\text{acf}}^p = T_{\text{acf}} \cup \{\phi_p\}$, for a prime number p, is the theory of algebraically closed fields of characteristic p.

$T_{\text{acf}}^0 = T_{\text{acf}} \cup \{\neg \phi_n \mid n > 0\}$ is the theory of algebraically closed fields of characteristic zero.

Then the theories T_{acf}^p and T_{acf}^0 are complete since, by quantifier elimination, we only have to look at quantifier-free sentences. These are combinations (using $\wedge$, $\vee$, $\neg$ and $\rightarrow$) of sentences $t = s$, with t and s closed terms. Then t and s represent elements of $\mathbb{Z}$, and $t = s$ is a consequence of T_{acf}^p precisely when their difference is a multiple of p. For characteristic 0 we have: the equality $t = s$ is a consequence of T_{acf}^0 precisely when this sentence is true in $\mathbb{Z}$.

The completeness of these theories has the following consequence. We write $\mathbb{F}_p$ for the field of p elements, and $\overline{\mathbb{F}_p}$ for its algebraic closure. $\mathbb{C}$ is the field of complex numbers.

Lemma 2.7.7 *Let ϕ be a sentence of L_{rings}. The following assertions are equivalent:*

(i) $\mathbb{C} \models \phi$.

(ii) *There is a natural number m such that for all primes $p > m$, $\overline{\mathbb{F}_p} \models \phi$.*

Proof. The field $\mathbb{C}$ is algebraically closed and of characteristic zero, so if $\mathbb{C} \models \phi$ then by completeness of T^0_{acf}, $T^0_{\text{acf}} \models \phi$. By the Compactness Theorem, there is a number m such that $T_{\text{acf}} \cup \{\neg\phi_n \mid n \leq m\} \models \phi$. It follows that for every $p > m$, $\overline{\mathbb{F}_p} \models \phi$. This proves (i)$\Rightarrow$(ii); the converse implication is proved in the same way, considering $\neg\phi$ instead of ϕ. ∎

The following theorem is a nice application of this lemma.

Theorem 2.7.8 *Let $F_1, \ldots, F_n$ be polynomials in n variables $Y_1, \ldots, Y_n$ and with complex coefficients. Consider the function $f : \mathbb{C}^n \to \mathbb{C}^n$ defined by*

$$f(z_1, \ldots, z_n) = (F_1(z_1, \ldots, z_n), \ldots, F_n(z_1, \ldots, z_n)).$$

Then if f is injective, it is also surjective.

Proof. Convince yourself that for every natural number $d > 0$ there exists an L_{rings}-sentence Φ_d which expresses: "for every n-tuple of polynomials of degree $\leq d$, if the associated function f of n variables is injective, then it is surjective".

For an application of Lemma 2.7.7, we show that Φ_d is true in every field $\overline{\mathbb{F}_p}$. Indeed, suppose we have n polynomials $F_1, \ldots, F_n$ of degree $\leq d$ and coefficients in $\overline{\mathbb{F}_p}$ such that the function $f : (\overline{\mathbb{F}_p})^n \to (\overline{\mathbb{F}_p})^n$ is injective. Consider an n-tuple $(a_1, \ldots, a_n) \in (\overline{\mathbb{F}_p})^n$. Let $b_1, \ldots, b_k$ be the list of coefficients which occur in the F_i. There is a least subfield F of $\overline{\mathbb{F}_p}$ which contains all a_i and all b_j. Then F is a finite extension of $\mathbb{F}_p$, hence finite. Moreover, f restricts to a function $F^n \to F^n$ which is still injective. But every injective function from a finite set to itself is also surjective. We conclude that $(a_1, \ldots, a_n)$ is in the image of f. Therefore, $\overline{\mathbb{F}_p} \models \Phi_d$.

By Lemma 2.7.7 we have $\mathbb{C} \models \Phi_d$, which was what we had to prove. ∎

Another application of quantifier elimination for algebraically closed fields concerns Hilbert's Nullstellensatz. We have to invoke a result from algebra.

Lemma 2.7.9 (Hilbert's Basis Theorem) *For every field K, every ideal of the polynomial ring $K[X_1, \ldots, X_n]$ is finitely generated.*

Proof. See, e.g., [47]. ∎

We can now prove the following part of Hilbert's classical Nullstellensatz (called the "weak form" in [11]).

Theorem 2.7.10 (Hilbert's Nullstellensatz, weak form) *Suppose k is an algebraically closed field and $k[X_1, \ldots, X_n]$ the polynomial ring over k in n variables. Suppose I is an ideal in $K[X_1, \ldots, X_n]$. Then either $1 \in I$ or there are elements $a_1, \ldots, a_n$ in k such that $g(a_1, \ldots, a_n) = 0$ for every $g \in I$.*

Proof. Suppose $1 \notin I$. Then I is contained in a maximal ideal M of $k[X_1, \ldots, X_n]$. Let k' be the algebraic closure of the quotient field

$$k[X_1, \ldots, X_n]/M.$$

In k', the elements $\overline{X_1}, \ldots, \overline{X_n}$ have the property that $g(\overline{X_1}, \ldots, \overline{X_n}) = 0$ for every $g \in I$. Here $\overline{X_i}$ denotes the residue class of X_i modulo M.

Now k is a subring of k' and both are algebraically closed fields; by quantifier elimination, k is an elementary substructure of k'. It follows that for any *finite* number $g_1, \ldots, g_m$ of elements of I,

$$k \models \exists y_1 \cdots y_n (g_1(\vec{y}) = 0 \wedge \cdots \wedge g_m(\vec{y}) = 0).$$

But by Hilbert's Basis Theorem (Lemma 2.7.9) every ideal of $k[X_1, \ldots, X_n]$ is finitely generated, so we are done.   ∎

There are many more applications of Logic (Model Theory) to Algebra. For a modern introduction to this area, see [51]. See also the Appendix.

2.8   The Löwenheim–Skolem Theorems

The theorems in this section concern the question of how "big" a model of a consistent first-order L-theory T can be. Of course, it can happen that T contains a sentence which forces every model of T to have cardinality $\leq n$ for some $n \in \mathbb{N}$, as we have seen. It is also possible that a theory forces models to be at least as big as a given set C: for example if L has constants for each element of C, and the theory has axioms

$$\neg(c = d)$$

for each pair (c, d) of distinct constants.

The upshot of this section will be that this is basically all a theory can say. If there is an infinite model of T, there will, in general, be models of T of every infinite cardinality greater than a certain cardinal number associated with the language L. The following result is due to Leopold Löwenheim (1878–1957) and Thoralf Skolem (1887–1963).

Theorem 2.8.1 (Upward Löwenheim–Skolem Theorem)   *Suppose T has an infinite model. Then for any set C there is a model M of T such that $|C| \leq |M|$.*

Proof. Let L_C be the language L of the theory T, together with new constants c for every $c \in C$. We consider the L_C-theory T_C, which has all the axioms of T, together with the axioms

$$\neg(c = d)$$

for every pair of distinct elements c, d of C.

If M is a model of T_C, then M is a model of T, and moreover, the assignment $c \mapsto c^M$ specifies a function from C into M, which is injective since $M \models \neg(c = d)$ (which means $c^M \neq d^M$) whenever $c \neq d$. So all we have to do is show that T_C is consistent. This is done with the Compactness Theorem.

Let $T' \subseteq T_C$ be a finite subtheory. Then in T', only finitely many constants from C occur, say $c_1, \ldots, c_n$. Now by assumption T has an infinite model N; take n distinct elements $a_1, \ldots, a_n$ from N and make N into an L_C-structure by putting $(c_i)^N = a_i$ for $i = 1, \ldots, n$ and, for $c \neq c_1, \ldots, c_n$, let c^N be an arbitrary element of N.

Then N is an L_C-structure which is a model of T'. Hence, every finite subtheory T' of T_C has a model; so T_C has a model by the Compactness Theorem. ∎

The proof can be refined to obtain "large" models of T with certain extra properties. For example, if N is an infinite model of T and C is any set, then there is a model M of T such that there is an injective function $C \to M$ and moreover, M satisfies exactly the same L-sentences as N.

Exercise 103 Prove this last statement. [Hint: instead of T, use the set of L-sentences which are true in N.]

The following corollary is a strengthening of Theorem 2.8.1.

Corollary 2.8.2 *Let N be an infinite model of a theory T, and C an arbitrary set. Then there exists a model M of T which contains N as an elementary substructure and allows an injective function: $C \to M$.*

Proof. Apply Theorem 2.8.1 to the theory $E(N)$ (see Exercise 96) and note that $T \subseteq E(N)$. ∎

We now state the "downward Löwenheim–Skolem Theorem". Its formulation uses the notion of *cardinality of the language L*, notation $\|L\|$, which is by definition the cardinality of the set of L-formulas. Since L is also defined as a set (the set of all constants, function symbols and relation symbols), we also have the 'ordinary cardinality' $|L|$; the following exercise compares the two.

Exercise 104 Show that $\|L\| = |L|$ if L is infinite, and that $\|L\| = \omega$ if L is finite.

Theorem 2.8.3 (Downward Löwenheim–Skolem Theorem) *Let M be an infinite model of a theory T in a language L, and let $C \subseteq M$ be a subset with $|C| \leq \|L\|$. Then there is an elementary substructure $N \preceq M$ which contains C as a subset, and has the property that $|N| \leq \|L\|$.*

Proof. Let $C \subseteq M$ be given. If $C = \emptyset$, pick an arbitrary element $a \in M$ and put $C_0 = \{a\}$; otherwise, let $C_0 = C$.

The submodel N will be constructed as the union of a chain of subsets of M of cardinality $\leq \|L\|$:

$$C = C_0 \subseteq C_1 \subseteq C_2 \subseteq \cdots$$

This chain is constructed inductively as follows. The set C_0 is given. Suppose we have constructed C_k (and it is part of the induction hypothesis that $|C_k| \leq \|L\|$). Let N_k be the substructure of M generated by C_k. Then $|N_k| \leq \|L\|$ by Exercise 95. Now for each L-formula of the form $\exists x \varphi$ with free variables $y_1, \ldots, y_n$ and each n-tuple $m_1, \ldots, m_n$ of elements of N_k such that $M \models \exists x \varphi(x, m_1, \ldots, m_n)$, choose an element m of M such that $M \models \varphi(m, m_1, \ldots, m_n)$. Let C_{k+1} be N_k together with all elements m so chosen. Since $|N_k| \leq \|L\|$ and there are only $\|L\|$ many L-formulas, $|C_{k+1}| \leq \|L\|$ too. This completes the construction of the chain.

Let N be the union $\bigcup_{i=0}^{\infty} C_i$. Then N is a substructure of M, because N contains c^M for every constant c of L (check!), and if f is an n-place function symbol of L and $m_1, \ldots, m_n \in N$, then for some k the elements $m_1, \ldots, m_n$ are already in C_k, so $f^M(m_1, \ldots, m_n) \in N_k \subseteq C_{k+1} \subseteq N$. And N is the union of a countable family of subsets of M of cardinality $\leq \|L\|$, so $|N| \leq \|L\|$ as desired.

It remains to prove that N is an elementary substructure of M. For this, we use the characterization given in Exercise 97. Suppose $\exists x \varphi(x, y_1, \ldots, y_n)$ is an L-formula and $m_1, \ldots, m_n \in N$ are such that $M \models \exists x \varphi(x, m_1, \ldots, m_n)$. Then there is a natural number k such that already $m_1, \ldots, m_n \in C_k$. By construction of C_{k+1}, there is an $m \in C_{k+1}$ such that $M \models \varphi(m, m_1, \ldots, m_n)$; this m is also an element of N. By Exercise 97, N is an elementary substructure of M.   ∎

We wrap up this section by putting together Corollary 2.8.2 and Theorem 2.8.3 to obtain the following useful conclusion:

Corollary 2.8.4 *Let M be an infinite model of an L-theory T, and let C be a set such that $|C| \geq \|L\|$. Then there is a model N of T such that $|N| = |C|$ and, moreover, N and M satisfy the same L-sentences.*

Proof. First we apply Corollary 2.8.2 to obtain a model M' which contains M as an elementary substructure and allows an embedding $C \to M'$. Then M and M' satisfy the same L-sentences; in particular, M' is infinite.

Next, consider the language L_C which has an extra constant for every element of C. The injective function $C \to M'$ makes M' into an L_C-structure. Clearly $|C| \leq \|L_C\|$. Applying Theorem 2.8.3 with L_C in the role of L, we see that M' contains an elementary substructure N with $C \subseteq N$ and $|N| = |C|$. Then M and N satisfy the same L-sentences.   ∎

2.9   Categorical Theories

Let us consider, as an example, the theory of k-vector spaces discussed in Section 2.4.3.

If k is a finite field, any two k-vector spaces of the same cardinality are isomorphic as k-vector spaces, for if $|V| = |W|$ then any basis for V and any basis for W must have the same cardinality (if B is a finite basis for V, then $|V| = |k|^{|B|}$;

if B is infinite, then $|V| = |B|$). And any bijection between bases extends uniquely to a k-linear map which is an isomorphism of k-vector spaces.

If k is infinite, this is no longer true: let $k = \mathbb{Q}$. The $\mathbb{Q}$-vector space $\mathbb{Q}[X]$ is countable and therefore of the same cardinality as $\mathbb{Q}$, but it has infinite dimension over $\mathbb{Q}$ and hence cannot be isomorphic to $\mathbb{Q}$ as a vector space over itself.

However, it is true (and follows by much the same reasoning as for finite k) that if $|V| = |W| > |k|$, then V and W are isomorphic as k-vector spaces.

Let L_k be the language of k-vector spaces, and let T_k^∞ be the theory of infinite k-vector spaces. That is, T_k^∞ has the axioms for a k-vector space together with all the sentences $\neg\phi_n$ from Example 2.5.2.

Theorem 2.9.1   *The theory T_k^∞ is complete.*

Proof. Suppose that $T_k^\infty \not\models \varphi$ and $T_k^\infty \not\models \neg\varphi$, for some L_k-sentence φ. Then there are infinite k-vector spaces V and W with $V \models \neg\varphi$ and $W \models \varphi$. But then, if C is any set such that $|C| > \|L_k\|$, Corollary 2.8.4 gives us k-vector spaces V' and W' such that:

(i) V' satisfies the same L_k-sentences as V;
(ii) W' satisfies the same L_k-sentences as W;
(iii) $|V'| = |W'| = |C|$.

Then as we have just argued, V' and W' must be isomorphic as k-vector spaces, yet $V' \models \neg\varphi$ by (i), and $W' \models \varphi$ by (ii). But this is clearly impossible, by Exercises 91 and 92. ∎

Exercise 105   For another proof of the fact that T_k^∞ is complete: prove that T_k^∞ has quantifier elimination.

Definition 2.9.2   Let κ be a cardinal number. An L-theory T is called κ-*categorical* if for every pair M, N of models of T of cardinality κ, there is an isomorphism between M and N.

As we have seen, the theory of infinite k-vector spaces is κ-categorical if $\kappa > |k|$.

The proof of the following theorem is a straightforward generalization of the argument above that the theory T_k^∞ must be complete; it is therefore left as an exercise.

Theorem 2.9.3 (Łoś–Vaught Test)   *Suppose T is an L-theory which only has infinite models, and suppose T is κ-categorical for some $\kappa \geq \|L\|$. Then T is complete.*

Exercise 106   Prove Theorem 2.9.3.

We conclude this chapter by giving an example of a theory which is ω-categorical; the theory of *dense linear orders without end-points*. In this example it is the technique of the proof, rather than the result, which is important. This technique is known as *Cantor's back-and-forth argument*.

Definition 2.9.4 The theory T_d of dense linear orders without end-points is formulated in a language with just one binary relation symbol $<$, and has the following axioms:

$$\forall x \neg (x < x) \text{ irreflexivity}$$
$$\forall xyz(x < y \wedge y < z \rightarrow x < z) \text{ transitivity}$$
$$\forall xy(x < y \vee x = y \vee y < x) \text{ linearity}$$
$$\forall xy(x < y \rightarrow \exists z(x < z \wedge z < y)) \text{ density}$$
$$\forall x \exists yz(y < x \wedge x < z) \text{ no end points}$$

Theorem 2.9.5 (Cantor) *The theory T_d is ω-categorical.*

Proof. We have to show that any two countably infinite models M and N of T_d are isomorphic.

Start by choosing enumerations $M = \{m_0, m_1, \ldots\}$ and $N = \{n_0, n_1, \ldots\}$ of M and N.

We shall construct an isomorphism $\beta : M \rightarrow N$ as the union of a chain of order-preserving bijective functions between finite sets:

$$
\begin{array}{ccccccc}
M_0 & \longrightarrow & M_1 & \longrightarrow & M_2 & & \cdots \\
\beta_0 \downarrow & & \beta_1 \downarrow & & \beta_2 \downarrow & & \\
N_0 & \longrightarrow & N_1 & \longrightarrow & N_2 & & \cdots
\end{array}
$$

such that the horizontal arrows are inclusions $M_k \subseteq M_{k+1}$, $N_k \subseteq N_{k+1}$, and β_k is the restriction of β_{k+1} to M_k. Moreover, we shall make sure that for each k we have $\{m_0, \ldots, m_k\} \subseteq M_k$ and $\{n_0, \ldots, n_k\} \subseteq N_k$, so at the end we obtain a bijective function from M to N.

Let $M_0 = \{m_0\}$, $N_0 = \{n_0\}$ and let β_0 be the unique bijection.

Suppose $\beta_k : M_k \rightarrow N_k$ has been constructed, as an order-preserving bijection. We construct M_{k+1}, N_{k+1} and β_{k+1} in two stages:

Stage 1. If $m_{k+1} \in M_k$, we do nothing in this stage and proceed to stage 2. If $m_{k+1} \notin M_k$ there are two possibilities:

- Either m_{k+1} lies below all elements of M_k, or above all these elements. In this case, we use the axiom "no end-points" to find an element $n \in N$ which has the same relative position with respect to N_k; we add m_{k+1} to M_k and n to N_k. Then we put $\beta_{k+1}(m_{k+1}) = n$.
- m_{k+1} lies somewhere between the elements of M_k. Then by the axiom "linearity" and the fact that M_k is finite, there is a greatest element $m_j \in M_k$ and a least $m_l \in M_k$ such that $m_j < m_{k+1} < m_l$. We use the axiom "density" to pick an element n of N with $\beta_k(m_j) < n < \beta_k(m_l)$; we add m_{k+1} to M_k and n to N_k. Then we put $\beta_{k+1}(m_{k+1}) = n$.

Stage 2. Here we do the symmetric thing with n_{k+1} and the inverse of the finite bijective function we obtained after Stage 1. After completing Stage 2 we let $\beta_{k+1} : M_{k+1} \to N_{k+1}$ be the union of β_k and what we have added in Stages 1 and 2.

This completes the construction of β_{k+1} and hence, inductively, of our chain of finite bijective order-preserving functions.   ∎

Exercise 107 Show that the theory T_d is complete.

Exercise 108 Use Lemma 2.7.4 to prove that the theory T_d has quantifier elimination.

Exercise 109 Show that the theory T_d is *not* 2^ω-categorical.

Exercise 110 Let $T_d(a, b)$ be the theory with the same axioms as T_d, but now formulated in a language with two extra constants a, b. Prove that $T_d(a, b)$ has exactly three non-isomorphic countable models.

Exercise 111 Give an example of a theory which has exactly four non-isomorphic countably infinite models.

Exercise 112 Use Theorem 2.9.5 for another proof that $\mathbb{R}$ is not countable.

Exercise 113 Let L be the language $\{R\}$, where R is a binary relation symbol. Let T be the L-theory whose models are precisely those L-structures (M, R^M) in which R^M is an equivalence relation on M with infinitely many equivalence classes, and every equivalence class is infinite.

(a) Give a system of axioms for T.
(b) Prove that T is ω-categorical.
(c) Prove that T has quantifier elimination.

Chapter 3
Proofs

In Chapter 2 we introduced languages and formulas as mathematical objects: formulas are just certain finite sequences of elements of a certain set. Given a specific structure for a language, such formulas become mathematical statements via the definition of truth in that structure.

In mathematical reasoning, one often observes that one statement "follows" from another, without reference to specific models or truth, as a purely "logical" inference. More generally, statements can be conjectures, assumptions or intermediate conclusions in a mathematical argument.

In this chapter we shall give a formal, abstract definition of a concept called "proof". A proof will be a finite object which has a number of *assumptions* which are formulas, and a *conclusion* which is a formula. Given a fixed language L, there will be a set of all proofs in L, and we shall be able to prove the *Completeness Theorem*:

> For a set Γ of L-sentences and an L-sentence ϕ, the relation $\Gamma \models \phi$ holds if and only if there exists a proof in L with conclusion ϕ and assumptions from the set Γ.

Recall that $\Gamma \models \phi$ was defined as: for every L-structure M which is a model of Γ, it holds that $M \models \phi$.

Therefore, the Completeness Theorem reduces a universal ("for all") statement about a large class of structures, to an existential ("there is") statement about one set (the set of proofs). Furthermore, we shall see that proofs are built up by rules that can be interpreted as elementary reasoning steps (we shall not go into the philosophical significance of this). Finally, we wish to remark that it can be effectively tested whether or not an object of the appropriate kind is a "proof", and that the set of all sentences ϕ such that $\Gamma \models \phi$ can be effectively generated by a computer (see Section A.1 in the Appendix for a precise interpretation of this).

Mathematicians who devised definitions of a notion of "formal proof" include Frege, Russell and Hilbert; but by far the most influential one is due to Gerhard

© Springer Nature Switzerland AG 2018
I. Moerdijk, J. van Oosten, *Sets, Models and Proofs*, Springer Undergraduate
Mathematics Series, https://doi.org/10.1007/978-3-319-92414-4_3

Kurt Gödel

Gentzen (1909–1945). In fact, Gentzen gave two widely used systems, of which we present the first one below. He called his system "natural deduction" (*Kalkül des natürlichen Schließens* [21]). For biographical information about Gentzen (who was a member of the SA and the Nazi Party, and died in a prison camp in 1945) see [54].

The Completeness Theorem (the cornerstone of first-order logic) was proved by Kurt Gödel in 1929 ([22]). Gödel was arguably the most important logician of the 20th century. For a biography, see [14]; for an overview of Gödel's work, see the book [83]. Our treatment in Section 3.2 is based on a proof by Leon Henkin (1921–2006) given in [32].

3.1 Proof Trees

In a well-structured mathematical argument, it is clear at every point what the conclusion reached so far is, what the current assumptions are and on which intermediate results each step depends.

We model this mathematically with the concept of a *tree*.

Definition 3.1.1 A *tree* is a partial order $(T, \leq)$ which has a least element, and is such that for every $x \in T$, the set

$$\downarrow(x) \equiv \{y \in T \mid y \leq x\}$$

is well-ordered by the relation $\leq$.

We shall only be concerned with *finite* trees; that is, finite posets T with least element, such that each $\downarrow(x)$ is linearly ordered.

This is an example of a tree:

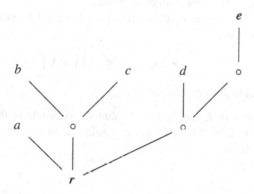

We use the following dendrological language when dealing with trees: the least element is called the *root* (in the example above, the element marked r), and the maximal elements are called the *leaves* (in the example, the elements marked a, b, c, d, e). The lines indicate the order (going upwards), and the o's are unnamed elements.

When we see a proof as a tree, the leaves are the places for the assumptions, and the root is the place for the conclusion. The information that the assumptions give may be compared to the carbon dioxide in real trees, which finds its way from the leaves to the root.

The following exercise gives some alternative ways of characterizing trees.

Exercise 114

(a) Show that a finite tree is the same thing as a finite sequence of nonempty finite sets and functions

$$A_n \rightarrow \cdots \rightarrow A_1 \rightarrow A_0$$

where A_0 is a one-element set.

(b) Show that a finite tree is the same thing as a finite set V together with a function $f : V \rightarrow V$ which has the properties that f has exactly one fixed point $r = f(r)$, and there are no elements $x \neq r$ such that $x = f^n(x)$ for some $n \in \mathbb{N}_{>0}$.

(c) If V is a finite set, a *hierarchy* on V is a collection $\mathcal{C}$ of subsets of V such that $V \in \mathcal{C}$, and for any two elements $C_1 \neq C_2$ of $\mathcal{C}$, we have $C_1 \subset C_2$ or $C_2 \subset C_1$

or $C_1 \cap C_2 = \emptyset$. Let us call a hierarchy $\mathcal{C}$ a T_0-*hierarchy* if for each $x, y \in V$ with $x \neq y$, there is a $C \in \mathcal{C}$ such that either $x \in C$ and $y \notin C$, or $y \in C$ and $x \notin C$. Call $\mathcal{C}$ *connected* if there is an element $r \in V$ such that the only element $C \in \mathcal{C}$ such that $r \in C$ is V itself.

(i) Show that if $\mathcal{C}$ is a connected T_0-hierarchy on V, then the relation

$$x \leq y \text{ if and only if for all } C \in \mathcal{C}, x \in C \text{ implies } y \in C$$

defines a partial order on V which is a tree; and moreover, for every $x \in V$, the set $\{y \in V \mid x \leq y\}$ is an element of $\mathcal{C}$.

(ii) Show also that if $\leq$ is a partial order on a finite set V which is a tree, then the set of subsets of V

$$\{\{y \in V \mid x \leq y\} \mid x \in V\}$$

is a connected T_0-hierarchy on V.

We shall be interested in L-*labelled* trees; that is: trees where the elements have 'names' which are L-formulas or formulas marked with a symbol † (pronounced "dagger"). For example:

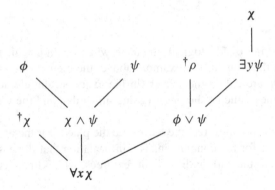

The following definition formalizes this.

Definition 3.1.2 Let L be a language. We fix an extra symbol †. A *marked L-formula* is a pair $(\dagger, \varphi)$; we shall write $^\dagger\varphi$ for $(\dagger, \varphi)$. Let $F(L)$ be the set of L-formulas, and let $^\dagger F(L)$ be the disjoint union of $F(L)$ and the set $\{\dagger\} \times F(L)$ of marked L-formulas.

An L-*labelled* tree is a finite tree T together with a function f from T to the set $^\dagger F(L)$ such that the only elements x of T such that $f(x)$ is a marked formula are leaves of T.

The function f is called the *labelling function*, and $f(x)$ is called the *label* of x.

Remark 3.1.3 In other contexts in mathematics the notion of "planar tree" is important. A *planar tree* is a tree together with a specific embedding of the tree into the plane. Every picture of a tree can be viewed as such an embedding. Our trees are *not* planar: we regard the pictures

as two embeddings of the same tree.

Moreover, we call two labelled trees *equivalent* if the underlying partial orders are isomorphic via an isomorphism which commutes with the labelling functions.

Among the L-labelled trees, we shall single out the set of *proof trees*. The definition (Definition 3.1.4 below) uses the following two operations on L-labelled trees:

(1) *Joining a number of labelled trees by adding a new root labelled ϕ*

Suppose we have a finite number of labelled trees $T_1, \ldots, T_k$ with labelling functions $f_1, \ldots, f_k$. Let T be the disjoint union $T_1 + \cdots + T_k$ together with a new element r, and ordered as follows: $x \leq y$ if and only if either $x = r$ or x and y belong to the same T_i and $x \leq y$ holds in T_i.

Let the labelling function f on T be such that it extends each f_i on T_i and has $f(r) = \phi$.

We denote this construction by $\Sigma(T_1, \ldots, T_k; \phi)$. A picture:

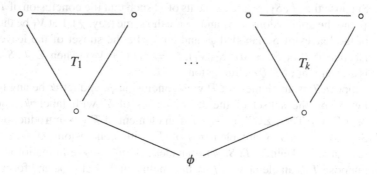

(2) *Adding some markings*

Suppose T is a labelled tree with labelling function f. If V is a set of leaves of T, we may modify f to f' as follows: $f'(x) = f(x)$ if $x \notin V$ or $f(x)$ is a marked formula; otherwise, $f'(x) = (\dagger, f(x))$.

We denote this construction by $Mk(T; V)$.

Exercise 115 Show that, up to equivalence of labelled trees (see Remark 3.1.3), every L-labelled tree can be constructed by a finite number of applications of these two constructions, starting from one-element trees with unmarked labels.

For the rest of this section, we shall assume that we have a fixed language L which we won't mention (we say 'labelled' and 'formula' instead of 'L-labelled', 'L-formula' etc.).

If T is a labelled tree with labelling function f, root r and leaves $a_1, \ldots, a_n$, we shall call the formula $f(r)$ (if it is a formula, that is: unmarked) the *conclusion* of T and the formulas $f(a_i)$ the *assumptions* of T. Assumptions of the form ${}^\dagger \varphi$ are called *eliminated assumptions*.

We can now give the promised definition of 'proof tree'. Instead of reading through the definition in one go, you are advised to work through a few clauses at a time and then have a look at the examples, referring back to the definition when necessary.

Definition 3.1.4 The set $\mathcal{P}$ of *proof trees* is the smallest set of labelled trees, satisfying:

Ass For every formula φ, the tree with one element r and labelling function $f(r) = \varphi$ is an element of $\mathcal{P}$. Note that φ is both an assumption and the conclusion of this tree. We call this tree an *assumption tree*.

$\wedge I$ If T_1 and T_2 are elements of $\mathcal{P}$ with conclusions φ_1 and φ_2, respectively, then $\Sigma(T_1, T_2; \varphi_1 \wedge \varphi_2)$ is an element of $\mathcal{P}$. We say this tree was formed by $\wedge$-*introduction*.

$\wedge E$ If T is an element of $\mathcal{P}$ with conclusion $\phi \wedge \psi$ then both $\Sigma(T; \phi)$ and $\Sigma(T; \psi)$ are elements of $\mathcal{P}$. These are said to be formed by $\wedge$-*elimination*.

$\vee I$ If T is an element of $\mathcal{P}$ with conclusion φ or ψ then $\Sigma(T; \varphi \vee \psi)$ is an element of $\mathcal{P}$. We say this tree is formed by $\vee$-introduction.

$\vee E$ Suppose that T, S_1, S_2 are elements of $\mathcal{P}$ such that the conclusion of T is $\varphi \vee \psi$ and the conclusions of S_1 and S_2 are the same (say, χ). Let V_1 be the subset of the leaves of S_1 labelled φ, and let V_2 be the subset of the leaves of S_2 labelled ψ. Let $S_1' = Mk(S_1; V_1)$, $S_2' = Mk(S_2; V_2)$. Then $\Sigma(T, S_1', S_2'; \chi)$ is an element of $\mathcal{P}$ ($\vee$-elimination).

$\rightarrow I$ Suppose T is an element of $\mathcal{P}$ with conclusion φ, and let ψ be any formula. Let V be the subset of the set of leaves of T with label ψ, and $T' = Mk(T; V)$. Then $\Sigma(T'; \psi \rightarrow \varphi)$ is an element of $\mathcal{P}$ ($\rightarrow$-introduction).

$\rightarrow E$ Suppose T and S are elements of $\mathcal{P}$ with conclusions $\varphi \rightarrow \psi$ and φ, respectively. Then $\Sigma(T, S; \psi)$ is an element of $\mathcal{P}$ ($\rightarrow$-elimination).

$\neg I$ Suppose T is an element of $\mathcal{P}$ with conclusion $\bot$. Let φ be any formula, and V be the subset of the set of leaves of T labelled φ. Let $T' = Mk(T; V)$. Then $\Sigma(T'; \neg\varphi)$ is an element of $\mathcal{P}$ ($\neg$-introduction).

$\neg E$ Suppose T and S are elements of $\mathcal{P}$ with conclusions φ and $\neg\varphi$, respectively. Then $\Sigma(T, S; \bot)$ is an element of $\mathcal{P}$ ($\neg$-elimination).

$\bot E$ Suppose T is an element of $\mathcal{P}$ with conclusion $\bot$. Let φ be any formula, and V the subset of the set of leaves of T labelled $\neg\varphi$. Let $T' = Mk(T; V)$. Then $\Sigma(T'; \varphi)$ is an element of $\mathcal{P}$ ($\bot$-elimination; one also hears *reductio ad absurdum* or *proof by contradiction*).

Subst Suppose T and S are elements of $\mathcal{P}$ such that the conclusion of T is $\varphi[t/x]$ and the conclusion of S is $(t = s)$. Suppose furthermore that the substitutions $\varphi[t/x]$ and $\varphi[s/x]$ are legitimate (see Definition 2.2.14). Then $\Sigma(T, S; \varphi[s/x])$ is an element of $\mathcal{P}$ (Substitution).

∀I Suppose T is an element of $\mathcal{P}$ with conclusion $\varphi[u/v]$, where u is a variable which does not occur in any unmarked assumption of T or in the formula $\forall v \varphi$ (and is not bound in φ). Then $\Sigma(T; \forall v \varphi)$ is an element of $\mathcal{P}$ (∀-introduction).

∀E Suppose T is an element of $\mathcal{P}$ with conclusion $\forall u \varphi$, and t is a term such that the substitution $\varphi[t/u]$ is legitimate. Then $\Sigma(T; \varphi[t/u])$ is an element of $\mathcal{P}$ (∀-elimination).

∃I Suppose T is an element of $\mathcal{P}$ with conclusion $\varphi[t/u]$, and suppose the substitution $\varphi[t/u]$ is legitimate. Then $\Sigma(T; \exists u \varphi)$ is an element of $\mathcal{P}$ (∃-introduction).

∃E Suppose T and S are elements of $\mathcal{P}$ with conclusions $\exists x \varphi$ and χ, respectively. Let u be a variable which doesn't occur in φ or χ, and is such that the only unmarked assumptions of S in which u occurs are of the form $\varphi[u/x]$. Let V be the set of leaves of S with label $\varphi[u/x]$, and $S' = Mk(S; V)$. Then $\Sigma(T, S'; \chi)$ is an element of $\mathcal{P}$ (∃-elimination).

Examples. The following labelled trees are proof trees. Convince yourself of this, and find out at which stage labels have been marked:

(a)

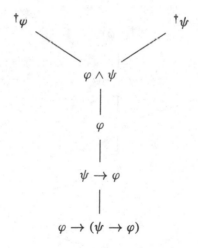

(b)

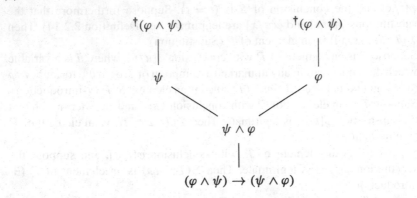

(c)

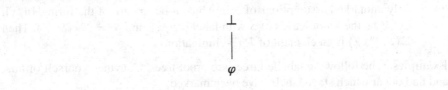

"Ex falso sequitur quodlibet"

(d)

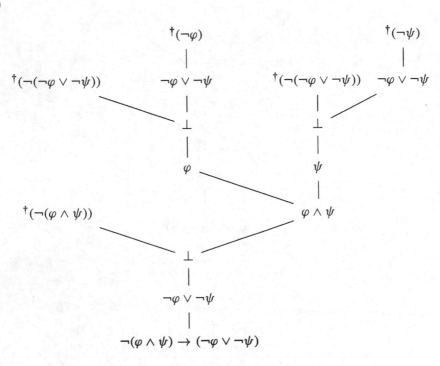

(e)

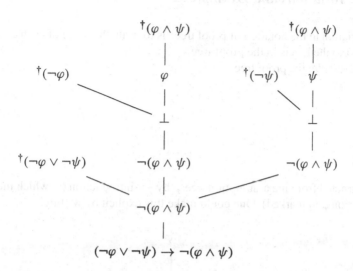

(f) The following example illustrates why, in formulating the rule $\forall I$, we have required that the variable u does not occur in the formula $\forall v\varphi$. For, let φ be the formula $u = v$. Consider that $(u = v)[u/v]$ is $u = u$, so were it not for this requirement, the following tree would be a valid proof tree:

$$\forall x(x = x)$$
$$|$$
$$u = u$$
$$|$$
$$\forall v(u = v)$$

Clearly, we would not like to accept this as a valid proof!

Definition 3.1.5 We define the relation

$$\Gamma \vdash \varphi$$

as: there is a proof tree with conclusion φ and whose unmarked assumptions are either elements of Γ or of the form $\forall x(x = x)$ for some variable x. We abbreviate $\{\varphi\} \vdash \psi$ as $\varphi \vdash \psi$, we write $\vdash \psi$ for $\emptyset \vdash \psi$, and $\Gamma, \varphi \vdash \psi$ for $\Gamma \cup \{\varphi\} \vdash \psi$.

Exercise 116 (Deduction Theorem) Prove that the statement $\Gamma, \varphi \vdash \psi$ is equivalent to $\Gamma \vdash \varphi \to \psi$.

3.1.1 Variations and Examples

One variation in the notation of proof trees is to write the name of each construction step next to the labels in the proof tree.

For example, the proof tree

is constructed from the assumption tree φ by $\to$-introduction (at which moment the assumption φ is marked). One could make this explicit by writing

Another notational variation is one that is common in the literature: the ordering is indicated by horizontal bars instead of vertical or skew lines, and next to these bars it is written by which of the constructions of Definition 3.1.4 the new tree results from the old one(s). Assumptions are numbered so that different assumptions have different numbers, but distinct occurrences of the same assumption may get the same number. The assumptions which are marked in the construction are indicated by their numbers next to the name of the construction. In this style, the proof tree

appears as follows:

$$\frac{^\dagger\varphi^1}{\varphi \to \varphi} \to I, 1$$

We shall call this a *decorated proof tree*. Although (or maybe: because!) they contain some redundant material, decorated proof trees are easier to read and better for practising the construction of proof trees.

In decorated style, examples (a)–(e) of the previous section are as follows:

(a)

$$\cfrac{\cfrac{\cfrac{\cfrac{^\dagger\varphi^1 \qquad ^\dagger\psi^2}{\varphi \wedge \psi}\wedge I}{\varphi}\wedge E}{\psi \to \varphi}\to I,2}{\varphi \to (\psi \to \varphi)}\to I,1$$

The assumption φ, numbered 1, gets marked when construction $\to I$ with number 1 is performed; etcetera.

(b)

$$\cfrac{\cfrac{\cfrac{^\dagger\varphi \wedge \psi^1}{\psi}\wedge E \qquad \cfrac{^\dagger\varphi \wedge \psi^1}{\varphi}\wedge E}{\psi \wedge \varphi}\wedge I}{(\varphi \wedge \psi) \to (\psi \wedge \varphi)}\to I,1$$

(c)

$$\cfrac{\bot}{\varphi}\bot E$$

(d)

$$\cfrac{\cfrac{^\dagger\neg(\varphi \wedge \psi)^4 \qquad \cfrac{\cfrac{^\dagger\neg(\neg\varphi \vee \neg\psi)^3 \quad \cfrac{^\dagger\neg\varphi^1}{\neg\varphi \vee \neg\psi}\vee I}{\bot}\neg E}{\varphi}\bot E,1 \qquad \cfrac{^\dagger\neg(\neg\varphi \vee \neg\psi)^3 \quad \cfrac{^\dagger\neg\psi^2}{\neg\varphi \vee \neg\psi}\vee I}{\bot}\neg E}{\bot}\bot E,2}{\varphi \wedge \psi}\wedge I}{\cfrac{\cfrac{\bot}{\neg\varphi \vee \neg\psi}\bot E,3}{\neg(\varphi \wedge \psi) \to (\neg\varphi \vee \neg\psi)}\to I,4}$$

(e)

$$\cfrac{^\dagger\neg\varphi \vee \neg\psi^5 \qquad \cfrac{\cfrac{^\dagger\neg\varphi^3 \quad \cfrac{^\dagger\varphi \wedge \psi^1}{\varphi}\wedge E}{\cfrac{\bot}{\neg(\varphi \wedge \psi)}\neg I,1}\neg E \qquad \cfrac{^\dagger\neg\psi^4 \quad \cfrac{^\dagger\varphi \wedge \psi^2}{\psi}\wedge E}{\cfrac{\bot}{\neg(\varphi \wedge \psi)}\neg I,2}\neg E}{\neg(\varphi \wedge \psi)}\vee E,3,4}{(\neg\varphi \vee \neg\psi) \to \neg(\varphi \wedge \psi)}\to I,5$$

Some more examples:

(f) A proof tree for $t = s \vdash s = t$:

$$\frac{\dfrac{\forall x(x = x)}{t = t} \, \forall E \qquad t = s}{s = t} \, \text{Subst}$$

The use of Subtitution is justified since $t = t$ is $(u = t)[t/u]$. Quite similarly, we have a proof tree for $\{t = s, s = r\} \vdash t = r$:

$$\frac{t = s \qquad s = r}{t = r} \, \text{Subst}$$

(g)

$$\frac{\dfrac{{}^{\dagger}\neg\exists x\varphi(x)^2 \qquad \dfrac{{}^{\dagger}\varphi(y)^1}{\exists x\varphi(x)} \, \exists I}{\dfrac{\bot}{\neg\varphi(y)} \, \neg I, 1}}{\dfrac{\forall x\neg\varphi(x)}{\neg\exists x\varphi(x) \to \forall x\neg\varphi(x)}} \to I, 2$$

You should check why the application of $\forall I$ is justified in this tree.

(h) The following tree gives an example of the $\exists E$-construction:

$$\frac{{}^{\dagger}\exists x\varphi(x)^2 \qquad \dfrac{\dfrac{\dfrac{{}^{\dagger}\forall x\neg\varphi(x)^3}{\neg\varphi(y)} \, \forall E \qquad {}^{\dagger}\varphi(y)^1}{\bot} \, \neg E}{\bot} \, \exists E, 1}{\dfrac{\dfrac{\bot}{\neg\exists x\varphi(x)} \, \neg I, 2}{\forall x\neg\varphi(x) \to \neg\exists x\varphi(x)}} \to I, 3$$

(i)

$$\frac{{}^{\dagger}\neg\forall x\neg\varphi(x)^3 \qquad \dfrac{\dfrac{{}^{\dagger}\neg\exists x\varphi(x)^2 \qquad \dfrac{{}^{\dagger}\varphi(y)^1}{\exists x\varphi(x)} \, \exists I}{\dfrac{\bot}{\neg\varphi(y)} \, \neg I, 1}}{\dfrac{\forall x\neg\varphi(x)}{} } \, \neg E}{\dfrac{\dfrac{\bot}{\exists x\varphi(x)} \, \bot E, 2}{\neg\forall x\neg\varphi(x) \to \exists x\varphi(x)}} \to I, 2$$

(j) The following tree is given in undecorated style; it is a good exercise to decorate it. It is assumed that the variables x and u do not occur in ϕ; check that without this condition, it is not a correct proof tree:

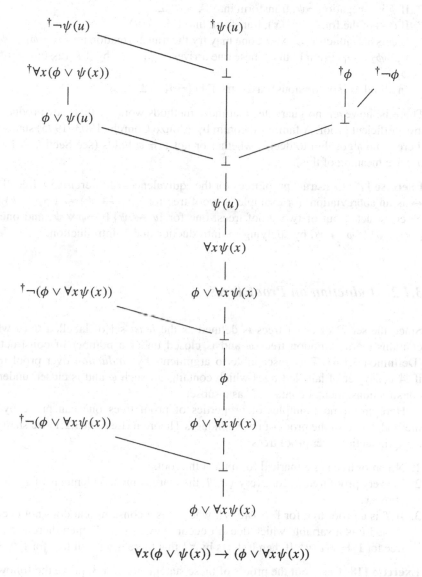

Remark 3.1.6 Some heuristics: when faced with the problem of constructing a proof tree which has a specified set of unmarked assumptions Γ and a prescribed conclusion ϕ (often formulated as: "construct a proof tree for $\Gamma \vdash \phi$"), it is advisable to use the following heuristics.

If ϕ is a conjunction $\phi_1 \wedge \phi_2$, break up the problem into two problems $\Gamma \vdash \phi_1$ and $\Gamma \vdash \phi_2$.

If ϕ is an implication $\phi_1 \to \phi_2$, transform the problem into $\Gamma, \phi_1 \vdash \phi_2$.

If ϕ is a negation $\neg \psi$, transform into $\Gamma, \psi \vdash \bot$.

If ϕ is of the form $\forall x \psi(x)$, transform into $\Gamma \vdash \psi(u)$.

If ϕ is a disjunction $\phi_1 \vee \phi_2$, one may try the transformation into $\Gamma, \neg\phi_1 \vdash \phi_2$ or $\Gamma, \neg\phi_2 \vdash \phi_1$. From both of these one arrives at $\phi_1 \vee \phi_2$ by $\bot E$ (assume $\neg(\phi_1 \vee \phi_2)$).

In all other (non-obvious) cases, try $\Gamma \cup \{\neg\phi\} \vdash \bot$.

There is, however, no guarantee that these methods work, or that they produce the most efficient proof. A famous theorem by Alonzo Church (1903–1995) states that there is no algorithm to decide whether or not $\Gamma \vdash \phi$ holds (see Section A.1 for a precise meaning of this).

Exercise 117 Construct proof trees for the equivalences of Exercise 59. Recall that $\leftrightarrow$ is an abbreviation: for example, a proof tree for $\vdash (\varphi \to \psi) \leftrightarrow (\neg\varphi \vee \psi)$ will be constructed out of two proof trees, one for $\{\varphi \to \psi\} \vdash \neg\varphi \vee \psi$, and one for $\{\neg\varphi \vee \psi\} \vdash \varphi \to \psi$, by applying $\to$-introduction and $\wedge$-introduction.

3.1.2   Induction on Proof Trees

Since the set $\mathcal{P}$ of proof trees is defined as the *least* set of labelled trees which contains the assumption trees φ and is closed under a number of constructions (Definition 3.1.4), $\mathcal{P}$ is susceptible to arguments by *induction* over proof trees: if $\mathcal{A}$ is any set of labelled trees which contains all such φ and is closed under the constructions, then $\mathcal{A}$ contains $\mathcal{P}$ as a subset.

Here are some examples of properties of proof trees one can prove by this method. Later, in the proof of the Soundness Theorem (Lemma 3.2.1) we shall also apply induction over proof trees.

1. No proof tree has a marked formula at the root.
2. In every proof tree T, for every $x \in T$ there are at most 3 elements of T directly above x.
3. If T is a proof tree for $\Gamma \vdash \varphi[c/u]$, where c is a constant that does not occur in Γ, and v is a variable which doesn't occur anywhere in T, then there is a proof tree for $\Gamma \vdash \varphi[v/u]$. It then follows by $\forall I$ that there is a proof tree for $\Gamma \vdash \forall u \varphi$.

Exercise 118 Carry out the proofs of these statements. For 3, prove the following: if T, c and v are as in the hypothesis, and $T[v/c]$ results from T by replacing c by v throughout, then $T[v/c]$ is also a proof tree, and is a proof tree for $\Gamma \vdash \varphi[v/u]$.

Exercise 119 Let $\Gamma \vdash_H \varphi$ be defined as the least relation between sets of L-formulas Γ and L-formulas φ such that the following conditions are satisfied:

(i) $\Gamma \vdash_H \forall x(x = x)$ always;

(ii)   if $\varphi \in \Gamma$, then $\Gamma \vdash_H \varphi$;

(iii) if $\Gamma \vdash_H \varphi$ and $\Gamma \vdash_H \psi$, then $\Gamma \vdash_H (\varphi \wedge \psi)$, and conversely;

(iv) if $\Gamma \vdash_H \varphi$ or $\Gamma \vdash_H \psi$, then $\Gamma \vdash_H (\varphi \vee \psi)$;

(v) if $\Gamma \cup \{\varphi\} \vdash_H \chi$ and $\Gamma \cup \{\psi\} \vdash_H \chi$, then $\Gamma \cup \{\varphi \vee \psi\} \vdash_H \chi$;

(vi) if $\Gamma \cup \{\varphi\} \vdash_H \perp$, then $\Gamma \vdash_H \neg\varphi$;

(vii) if $\Gamma \vdash_H \varphi$ and $\Gamma \vdash_H \neg\varphi$, then $\Gamma \vdash_H \perp$;

(viii) if $\Gamma \cup \{\neg\varphi\} \vdash_H \perp$, then $\Gamma \vdash_H \varphi$;

(ix) if $\Gamma \cup \{\varphi\} \vdash_H \psi$, then $\Gamma \vdash_H \varphi \rightarrow \psi$;

(x) if $\Gamma \vdash_H \varphi$ and $\Gamma \vdash_H \varphi \rightarrow \psi$, then $\Gamma \vdash_H \psi$;

(xi) if $\Gamma \vdash_H \psi(u)$ and u does not occur in Γ or in $\forall x \psi(x)$, then $\Gamma \vdash_H \forall x \psi(x)$;

(xii) if $\Gamma \vdash_H \forall x \psi(x)$ and the substitution $\psi[t/x]$ is legitimate, then $\Gamma \vdash_H \psi[t/x]$;

(xiii) if $\Gamma \vdash_H \psi[t/x]$ and the substitution $\psi[t/x]$ is legitimate, then $\Gamma \vdash_H \exists x \psi(x)$;

(xiv) if $\Gamma \cup \{\psi(u)\} \vdash_H \chi$ and u does not occur in Γ, χ or $\exists x \psi(x)$, then $\Gamma \cup \{\exists x \psi(x)\} \vdash_H \chi$;

(xv) if $\Gamma \vdash_H \varphi[t/x]$ and $\Gamma \vdash_H t = s$ while the substitutions $\varphi[s/x]$ and $\varphi[t/x]$ are legitimate, then $\Gamma \vdash_H \varphi[s/x]$.

Show that the relation $\Gamma \vdash_H \varphi$ coincides with the relation $\Gamma \vdash \varphi$ from Definition 3.1.5.

3.2 Soundness and Completeness

We compare the relation $\Gamma \vdash \phi$ from Definition 3.1.5 to the relation $\Gamma \models \phi$ from Chapter 2; recall that the latter means: in every model M of Γ, the sentence ϕ holds.

In this section we shall prove that the statements $\Gamma \models \phi$ and $\Gamma \vdash \phi$ are equivalent. This is the Completeness Theorem as announced in the beginning of this chapter. The Completeness Theorem says two important things about our proof system in relation to our notion of truth: our proof system is

sound, that is: whatever can be proved is always true;

complete, that is: whatever is always true can be proved.

Often, the term "Completeness Theorem" is used for the second of these statements, which is also the harder of the two.

Lemma 3.2.1 *Suppose T is an L-labelled proof tree with unmarked assumptions $\varphi_1, \ldots, \varphi_n$ and conclusion ψ; let $u_1, \ldots, u_k$ be a list of all variables that are free in at least one of $\varphi_1, \ldots, \varphi_n, \psi$. Then for every L-structure M and any k-tuple $m_1, \ldots, m_k$ of elements of M, we have:*

$$\text{If for all } i, \ 1 \leq i \leq n, \ M \models \varphi_i[m_1/u_1, \ldots, m_k/u_k], \text{ then}$$

$$M \models \psi[m_1/u_1, \ldots, m_k/u_k].$$

Proof. Lemma 3.2.1 is proved by a straightforward induction on proof trees: let $\mathcal{A}$ be the set of L-labelled proof trees which satisfy the condition of the lemma for every L-structure M.

Clearly, $\mathcal{A}$ contains every assumption tree φ. Now we need to show that $\mathcal{A}$ is closed under all the constructions of Definition 3.1.4. In most cases, a quick inspection suffices. We shall treat a few cases, leaving the others for you to check.

Let us write $\varphi_i[\vec{m}/\vec{u}]$ for $\varphi_i[m_1/u_1, \ldots, m_k/u_k]$.

Suppose T is formed by $\rightarrow$-introduction from $S \in \mathcal{A}$; say S has conclusion ψ and T has conclusion $\phi \rightarrow \psi$. Suppose the unmarked assumptions of S other than ϕ are $\varphi_1, \ldots, \varphi_n$, and let $u_1, \ldots, u_k$ be a list of variables as in the lemma, for S. Then if M is an L-structure and $m_1, \ldots, m_k \in M$, the induction hypothesis (viz, $S \in \mathcal{A}$) gives us that if $M \models \phi[\vec{m}/\vec{u}]$ and $M \models \varphi_i[\vec{m}/\vec{u}]$ for all $i \le n$ then $M \models \psi[\vec{m}/\vec{u}]$. Now clearly, if $M \models \varphi_i[\vec{m}/\vec{u}]$ for each $i \le n$, also $M \models (\phi \rightarrow \psi)[\vec{m}/\vec{u}]$. So $T \in \mathcal{A}$.

Suppose T is formed by $\forall$-introduction from $S \in \mathcal{A}$. Suppose S has unmarked assumptions $\varphi_1, \ldots, \varphi_n$ and conclusion $\psi(v)$, and v does not occur in $\varphi_1, \ldots, \varphi_n$. The induction hypothesis gives us that for any L-structure M and any tuple $\vec{m}$, p from M, if $M \models \varphi_i[\vec{m}/\vec{u}]$ for each $i \le n$ then $M \models \psi[\vec{m}/\vec{u}, p/v]$. Therefore, if $M \models \varphi_i[\vec{m}/\vec{u}]$ for each $i \le n$, then $M \models \psi[\vec{m}/\vec{u}, p/v]$ for all $p \in M$. In other words, $M \models (\forall x \psi[x/v])[\vec{m}/\vec{u}]$. Hence $T \in \mathcal{A}$.

Suppose T is formed by $\exists$-elimination from elements S, S' of $\mathcal{A}$. So the conclusion of S is $\exists v \phi$, the conclusion of S' is χ, and S' has possibly unmarked assumptions of the form $\phi[w/x]$, where w does not occur in any other unmarked assumption of S', nor in ϕ, nor in χ. Let $\vec{u}$ be the list of free variables appearing in an unmarked assumption of T or in χ. Let $\vec{m}$ be a tuple of elements of M of the same length as $\vec{u}$.

Suppose that $M \models \varphi[\vec{m}/\vec{u}]$ for each unmarked assumption φ of T. We need to show that $M \models \chi[\vec{m}]$. A little care is needed, for when we wish to apply the induction hypothesis to the trees S and S' we face the apparent problem that the formula $\exists v \phi$, which need not occur as an unmarked assumption or as the conclusion of T, may contain free variables $\vec{y}$ not among the $\vec{u}$. So let's write $\exists v \phi(\vec{u}, \vec{y}, v)$, displaying all the variables. Now the induction hypothesis for S tells us that for every tuple $\vec{n}$ of M of the same length as $\vec{y}$, $M \models \exists v \phi(\vec{m}, \vec{n}, v)$. Since M is nonempty, just pick any such tuple $\vec{n}_0$ from M. Then, choose $a \in M$ such that $M \models \phi(\vec{m}, \vec{n}_0, a)$. Now the induction hypothesis for S' (with the tuple $\vec{m}, \vec{n}_0, a$ for the variables $\vec{u}, \vec{y}, w$) tells us that $M \models \chi(\vec{m})$, as desired. $\blacksquare$

Exercise 120 Supply the induction step for the case of $\vee$-elimination, in the proof above.

Corollary 3.2.2 (Soundness Theorem) *Let Γ be an L-theory and ϕ an L-sentence. If $\Gamma \vdash \phi$ then $\Gamma \models \phi$.*

Proof. Immediate from Lemma 3.2.1. $\blacksquare$

Definition 3.2.3

(i) An L-theory Γ is called *formally consistent* if $\Gamma \nvdash \bot$.

(ii) An L-theory Γ is called *maximally formally consistent* if Γ is formally consistent but no proper extension of Γ is.

(iii) An L-theory is said to *have enough constants* if for every L-formula $\varphi(x)$ with one free variable x, there is a constant c such that

$$\Gamma \vdash \exists x \varphi(x) \to \varphi(c).$$

Exercise 121

(a) Show that the statement $\Gamma \nvdash \phi$ is equivalent to the statement that $\Gamma \cup \{\neg \phi\}$ is formally consistent.

(b) Suppose Γ is a maximally formally consistent set of L-sentences. Show that for any two L-sentences ϕ and ψ it holds that $\Gamma \vdash \phi \vee \psi$ if and only if either $\Gamma \vdash \phi$ or $\Gamma \vdash \psi$.

(c) Suppose Γ is maximally formally consistent. Prove that for any L-sentence ϕ, either $\phi \in \Gamma$ or $\neg \phi \in \Gamma$.

Lemma 3.2.4 *Let Γ be a maximally formally consistent set of L-sentences such that Γ has enough constants. Then Γ has a model.*

Proof. Let C be the set of constants of L. Then $C \neq \emptyset$ (why?). We define an equivalence relation $\sim$ on C by:

$$c \sim d \text{ if and only if } \Gamma \vdash (c = d).$$

It is easily verified (see Example (f) of Section 3.1.1) that $\sim$ is an equivalence relation. The set $M = C/\sim$ of equivalence classes is made into an L-structure as follows.

If F is an n-ary function symbol of L and $c_1, \ldots, c_n$ are elements of C, then $\Gamma \vdash \exists x (F(c_1, \ldots, c_n) = x)$; since Γ has enough constants, there is a constant c such that $\Gamma \vdash F(c_1, \ldots, c_n) = c$; define F^M by $F^M([c_1], \ldots, [c_n]) = [c]$. This is independent of the choices for c and the representatives $c_1, \ldots, c_n$, for if $c_i \sim d_i$ for $i = 1, \ldots, n$ and $c \sim d$, we have easily $\Gamma \vdash F(d_1, \ldots, d_n) = d$ by a number of Substitution constructions on the corresponding proof trees.

Similarly, if R is an n-place relation symbol we put

$$R^M = \{([c_1], \ldots, [c_n]) \mid \Gamma \vdash R(c_1, \ldots, c_n)\}$$

and again one checks that this is well-defined.

Finally, if c is a constant of L we let $c^M = [c]$. This completes the definition of M as an L-structure.

Now let t be a closed L-term. It is easily seen by induction on t that if c is a constant such that $\Gamma \vdash (t = c)$ (and such a constant exists, since Γ has enough constants), then $t^M = [c]$. Therefore, if s and t are closed L-terms, we have:

$$M \models (t = s) \text{ if and only if } \Gamma \vdash (t = s).$$

We shall now prove that for *every* L-sentence ϕ,

$$M \models \phi \text{ if and only if } \Gamma \vdash \phi$$

by induction on ϕ.

If ϕ is $R(c_1, \ldots, c_n)$, this holds by definition. Hence, since every closed term is equal to some constant, as we have just seen, the claim also holds for sentences $R(t_1, \ldots, t_n)$ where the t_i are closed terms.

Suppose ϕ is $\psi \vee \chi$. Then $M \models \phi$ if and only if (by definition of $\models$) $M \models \psi$ or $M \models \chi$, if and only if (by induction hypothesis) $\Gamma \vdash \psi$ or $\Gamma \vdash \chi$, if and only if (by Exercise 121, since Γ is maximally formally consistent) $\Gamma \vdash \psi \vee \chi$.

The step for $\neg\psi$ is similar, and the steps for $\wedge$ and $\rightarrow$ are left to you.

Now suppose ϕ is $\forall x \psi(x)$. We see that $M \models \phi$ is equivalent to: for all constants c of L, $M \models \psi(c)$. By induction hypothesis, this is equivalent to: for all constants c of L, $\Gamma \vdash \psi(c)$. This obviously follows from $\Gamma \vdash \forall x \psi(x)$. For the converse, using that Γ has enough constants, pick a c such that $\Gamma \vdash \exists x \neg\psi(x) \rightarrow \neg\psi(c)$. Then since $\Gamma \vdash \psi(c)$, we must have $\Gamma \vdash \neg\exists x \neg\psi(x)$. By one of the items of Exercise 117, $\Gamma \vdash \forall x \psi(x)$.

Again, the case for ϕ of the form $\exists x \psi(x)$ is similar, and is omitted.

We see that M is a model of Γ, which was to be proved.   ∎

Lemma 3.2.5 *Let Γ be a formally consistent set of L-sentences. Then there is an extension L' of L by constants, and a set Δ of L'-sentences which extends Γ and moreover is maximally formally consistent and has enough constants.*

Proof. Fix a set C which is disjoint from C_L^* (recall that this is the set of all symbols which can occur in L-formulas) and moreover satisfies $|C| = \|L\| = \max(\omega, |L|)$. Then C is infinite, so by Exercise 26 (a), $|C| = \omega \times |C|$; therefore, we can write C as a disjoint union:

$$C = \bigcup_{n \in \mathbb{N}} C_n$$

such that for each $n \in \mathbb{N}$, $|C_n| = |C|$.

Let L_0 be L, and $L_{n+1} = L_n \cup C_n$, where the elements of C_n are new constants. By induction, one sees that $\|L_n\| = \|L\| = |C|$. It follows, that for each n, there is an injective function from the set

$$F_n = \{\varphi(x) \mid \varphi(x) \text{ is an } L_n\text{-formula with one free variable } x\}$$

to the set C_n; we denote this map by $\varphi(x) \mapsto c_{\varphi(x)}$.

We let L' be $\bigcup_{n \in \mathbb{N}} L_n$. We construct Γ' as $\bigcup_{n \in \mathbb{N}} \Gamma_n$, where $\Gamma_0 = \Gamma$ and

$$\Gamma_{n+1} = \Gamma_n \cup \{\exists x \varphi(x) \rightarrow \varphi(c_{\varphi(x)}) \mid \varphi(x) \in F_n\}.$$

First, we prove the following fact:

(*) If $\Gamma_{n+1} \vdash \phi$, where ϕ is an L_n-sentence, then also $\Gamma_n \vdash \phi$.

Since every proof tree has only finitely many assumptions, we see that if $\Gamma_{n+1} \vdash \phi$ there are $\varphi_1(x), \ldots, \varphi_m(x) \in F_n$ such that

$$\Gamma_n \cup \{\exists x \varphi_i(x) \rightarrow \varphi_i(c_{\varphi_i(x)}) \mid 1 \leq i \leq m\} \vdash \phi.$$

Combining Exercise 116 and the equivalences of Exercise 117, one checks that this is equivalent to:

$$\Gamma_n \vdash \bigvee_{i=1}^{m} \neg(\exists x \varphi_i(x) \rightarrow \varphi_i(c_{\varphi_i(x)})) \vee \phi.$$

Now the constants $c_{\varphi_i(x)}$ are not in L_n, hence don't occur in Γ_n or in ϕ. It follows from Section 3.1.2 that

$$\Gamma_n \vdash \forall u_1 \cdots u_m [(\bigvee_{i=1}^{m} \neg(\exists x \varphi_i(x) \rightarrow \varphi_i(u_i))) \vee \phi].$$

By repeated use of $\vdash \forall x (\chi \vee \psi(x)) \rightarrow (\chi \vee \forall x \psi(x))$ (see Example (j) of Section 3.1.1), and $\vdash \neg(\alpha \rightarrow \beta) \rightarrow (\alpha \wedge \neg \beta)$,

$$\Gamma_n \vdash \bigvee_{i=1}^{m} (\exists x \varphi_i(x) \wedge \forall u_i \neg \varphi_i(u_i)) \vee \phi.$$

It follows, since $\vdash (\exists x \varphi_i(x) \wedge \forall u_i \neg \varphi_i(u_i)) \rightarrow \bot$ (check!), that $\Gamma_n \vdash \bot \vee \phi$, hence $\Gamma_n \vdash \phi$. This proves (*).

From (*) it follows that Γ' is formally consistent. For suppose $\Gamma' \vdash \bot$. Again using that every proof tree is finite, one finds that already $\Gamma_n \vdash \bot$ for some n; then by induction, using (*) one finds that $\Gamma \vdash \bot$, which contradicts the assumption that Γ is formally consistent.

It is easy to see that Γ' has enough constants; every formula contains only finitely many constants, so every L'-formula is an L_n-formula for some n. So a required constant for it will be in L_{n+1} by construction.

Now clearly, if a set of sentences has enough constants, then every bigger set also has enough constants. Therefore it suffices to show that Γ' can be extended to a maximally formally consistent set of L'-sentences; this is done with the help of Zorn's Lemma (Definition 1.3.4). Let P be the set of those sets of L'-sentences that contain Γ' and are formally consistent; P is ordered by inclusion. P is nonempty since $\Gamma' \in P$, as we have seen. If $\mathcal{K}$ is a chain in P then $\bigcup \mathcal{K}$ is formally consistent. Indeed, if $\bigcup \mathcal{K} \vdash \bot$ then already $Z \vdash \bot$ for some $Z \in \mathcal{K}$ (compare with the proof above that Γ' is formally consistent). By Zorn's Lemma, P has a maximal element Δ. Then Δ is maximally formally consistent, as is left for you to check; which finishes the proof. ∎

Corollary 3.2.6 (Completeness Theorem) *Let Γ be an L-theory and ϕ an L-sentence. If $\Gamma \models \phi$ then $\Gamma \vdash \phi$.*

Proof. We argue by contraposition. Suppose $\Gamma \nvdash \phi$. Then L-theory $\Gamma \cup \{\neg\phi\}$ is formally consistent. By Lemma 3.2.5 there is an extension L' of the language L and an L'-theory Δ such that $\Gamma \cup \{\neg\phi\} \subseteq \Delta$ and moreover Δ is maximally formally consistent and has enough constants. By Lemma 3.2.4 then, Δ has a model M. Now M is an L-structure which is a model of $\Gamma \cup \{\neg\phi\}$. It follows that $\Gamma \nvDash \phi$.   ∎

We can now give the promised proof of the Compactness Theorem from the Completeness Theorem (a proof which is independent of the theory of formal proofs was given in Section 2.5.1).

Corollary 3.2.7 (Compactness Theorem (2.5.1)) *If Γ is a set of sentences in a given language, and every finite subset of Γ has a model, then Γ has a model.*

Proof. Suppose Γ does not have a model. Then $\Gamma \vDash \bot$ so by the Completeness Theorem we have $\Gamma \vdash \bot$. Then, since proofs are finite, as we have already used a few times before, already $\Gamma' \vdash \bot$ for some finite $\Gamma' \subseteq \Gamma$. By the Soundness Theorem, $\Gamma' \vDash \bot$. This means that Γ' does not have a model, which contradicts our assumption that every finite subset of Γ has a model.   ∎

Exercise 122 Show that our proof of the Completeness Theorem has the corollary that every formally consistent set of L-sentences has a model of cardinality at most $\|L\|$. Compare this to Theorem 2.8.3.

3.3   Skolem Functions

Definition 3.3.1 Let L be a language. A *Skolem Theory* is an L-theory Δ with the property that for every L-formula $\varphi(\vec{x}, y)$ with $n + 1$ free variables, there is a function symbol F such that

$$\Delta \vdash \forall\vec{x}\, (\exists y \varphi(\vec{x}, y) \to \varphi(\vec{x}, F(\vec{x}))).$$

In the case $n = 0$, we take this to mean that for $\varphi(y)$ there is a constant c such that $\Delta \vdash \exists y \varphi(y) \to \varphi(c)$. So in particular, a Skolem theory has enough constants.

Definition 3.3.2 Suppose we have two languages $L \subseteq L'$ and two theories $T \subseteq T'$ such that T is an L-theory and T' is an L'-theory. Then T' is said to be *conservative over* T if every L-sentence ϕ which is a consequence of T' is already a consequence of T. In other words, $T' \vdash \phi$ implies $T \vdash \phi$ for every L-sentence ϕ.

Note that this notion already occurred in the proof of Lemma 3.2.5 where (*) states that Γ_{n+1} is conservative over Γ_n.

Exercise 123 Suppose that we have an infinite chain

$$L_1 \subseteq L_2 \subseteq \cdots$$

of languages, and also a chain

$$T_1 \subseteq T_2 \subseteq \cdots$$

of theories, such that for each n, T_n is an L_n-theory. Let $L = \bigcup_{n \geq 1} L_n$, and $T = \bigcup_{n \geq 1} T_n$. Then T is an L-theory. Prove, that if T_{n+1} is conservative over T_n for each $n \geq 1$, then T is conservative over T_1.

Theorem 3.3.3 *Let Γ be an L-theory. Then there is an extension L' of L, and a Skolem theory Δ in L' extending Γ, which is conservative over Γ.*

Proof. First, we show the following: for every L-theory Γ there is an extension L' of L and an L'-theory Δ such that $\Gamma \subseteq \Delta$ and Δ is conservative over Γ while for every L-formula $\varphi(\vec{x}, y)$ with $n + 1$ free variables, there is a function symbol F in L' such that

$$\Delta \vdash \forall \vec{x}\, (\exists y \varphi(\vec{x}, y) \rightarrow \varphi(\vec{x}, F(\vec{x}))).$$

Let L' be the extension of L obtained in the following way: for every L formula ψ and every string $(x_1, \ldots, x_k, y) = (\vec{x}, y)$ of variables such that all free variables of φ occur in $\vec{x}, y$, add a k-ary function symbol $F^{\varphi}_{\vec{x}, y}$. Let Δ be the set of L'-sentences defined as the union of Γ and the set of all L'-sentences of the form

$$\forall \vec{x}(\exists y \varphi \rightarrow \varphi[F^{\varphi}_{\vec{x}, y}(\vec{x})/y]),$$

where φ is an L-formula and $(\vec{x}, y)$ as above (the set Δ is said to be an extension of Γ by *Skolem functions*).

Exercise 124 Show that every model of Γ can be made into an L'-structure which is a model of Δ, by choosing appropriate functions as interpretations for the $F^{\varphi}_{\vec{x}, y}$. Then use the Completeness Theorem to conclude that Δ is conservative over Γ.

In order to prove the theorem, we iterate this construction infinitely often: let $L_1 = L$, and $T_1 = \Gamma$. Suppose L_n and T_n have been defined; let L_{n+1} and T_{n+1} then be the extended language and the extended theory which are obtained from L_n and T_n by the construction above.

Finally, let $L' = \bigcup_{n \geq 1} L_n$ and $\Delta = \bigcup_{n \geq 1} T_n$. By Exercise 123, Δ is conservative over Γ. The proof that Δ is a Skolem theory is left to you as the following exercise. ∎

Exercise 125 Finish the proof of Theorem 3.3.3: prove that the constructed theory Δ is in fact a Skolem theory.

Exercise 126 Let Δ be a Skolem theory, $M \models \Delta$. Show that every substructure of M is an elementary substructure; in particular, Δ has quantifier elimination.

Exercise 127 Let $L = \{R, f\}$ where R is a binary relation symbol and f a unary function symbol. The L-theory T has the following axioms:

> Axioms saying that R is an equivalence relation;
> $\forall x\, R(x, f(x))$;
> $\forall xy(R(x, y) \rightarrow f(x) = f(y))$.

(a) Let T_R be the theory in the language $\{R\}$ with only the axioms i). Is T conservative over T_R? Justify your answer.
(b) Let T_f be the theory in the language $\{f\}$ with no axioms. Is T conservative over T_f? Again, justify your answer.

Exercise 128 (This exercise was suggested to us by Jetze Zoethout.) In Exercise 124 we had an L-theory Γ, an extension of languages $L \subseteq L'$ and an L'-theory Δ extending Γ. We proved that Δ was conservative over Γ by showing that every model of Γ could be expanded (by finding suitable interpretations for the symbols in $L' - L$) to an L'-structure which was a model of Δ. Let us say that Δ is *strongly conservative* over Γ if this is the case. This exercise shows that not every conservative extension is strongly conservative.

Let L consist of just one binary relation symbol R and let Γ be the L-theory which has for each natural number $n \geq 2$ an axiom

$$\exists x_1 \cdots \exists x_n (R(x_1, x_2) \wedge R(x_2, x_3) \wedge \cdots \wedge R(x_{n-1}, x_n)).$$

Let L' extend L by a set of constants $c_0, c_1, \ldots$ and let Δ be the L'-theory which extends Γ by the axioms

$$\{R(c_i, c_{i+1}) \mid i \geq 0\}.$$

(a) Show that Δ is conservative over Γ. [Hint: use the Compactness Theorem.]
(b) Show that Δ is not strongly conservative over Γ. [Hint: consider an infinite well-order, and interpret R as $>$.]

Exercise 129 Let L be a language, T an L-theory and L' an extension of the language L.

(a) Show that the poset of all L'-theories which are conservative extensions of T, ordered by inclusion, satisfies the hypothesis of Zorn's Lemma.
(b) By Zorn's Lemma, there is a maximal L'-theory U which is conservative over T. Show that for every L'-sentence $\psi \notin U$, there are an L-sentence ϕ and an L'-sentence $\gamma \in U$ such that $\gamma \wedge \psi \models \phi$ and $T \not\models \phi$.

Chapter 4
Sets Again

This short chapter aims to give you a nodding acquaintance with the *formal theory of sets*. Most mathematicians accept this theory as a foundation for mathematics (see Remark 4.1.1 for what this might mean).

Set theory, as we saw in the introduction to Chapter 1, started with Cantor. We have already seen many of Cantor's results in this book: the Schröder–Cantor–Bernstein Theorem, the uncountability of $\mathbb{R}$, the diagonal argument, the Cantor Set, the notion of cardinal number and the Continuum Hypothesis, and in Chapter 2 the ω-categoricity of the theory of dense linear orders. The notion of "ordinal number", which we shall see in this chapter, is also due to him, and there is lots more.

Cantor was not a logician, and from a twentieth and twenty-first centuries perspective his theory was rather informal. Basically, a set could be formed by grouping together all objects sharing a certain property. This approach was also taken by Frege, who had a notion similar to Cantor's. However, Frege was one of the pioneers of modern logic and he added formal definitions for his constructions. It was precisely this that made his work susceptible to paradoxes, and one of these was pinpointed by Bertrand Russell in 1903 (this is the "antinomy" we alluded to in the introduction to Chapter 2). Consider the set $\mathbb{N}$ of natural numbers. Clearly, $\mathbb{N}$ is not a natural number, so it is not an element of $\mathbb{N}$. In symbols: $\mathbb{N} \notin \mathbb{N}$. Now, Russell continued, let us 'group together' into a set *all* those sets which are not elements of themselves: let

$$R = \{x \mid x \notin x\}.$$

© Springer Nature Switzerland AG 2018
I. Moerdijk, J. van Oosten, *Sets, Models and Proofs*, Springer Undergraduate
Mathematics Series, https://doi.org/10.1007/978-3-319-92414-4_4

Ernst Zermelo

Suppose R is a set. Then the question as to whether or not $R \in R$ makes sense. But by definition of R we find that $R \in R$ precisely when $R \notin R$! This is clearly a contradiction, which is known as "Russell's Paradox".[1]

4.1 The Axioms of ZF(C)

Mindful of Russell's paradox, Ernst Zermelo, whom we know from the Axiom of Choice and the Well-Ordering Theorem, took some care in formulating a system of axioms for sets in [86] (1908). One of the basic ideas is that one can group together all objects *from a given set* which have a certain property, to form a new set: instead of allowing $\{x \mid P(x)\}$ to be a set, we declare that $\{x \in X \mid P(x)\}$ is always a set provided X is a set (this is the Axiom Scheme of Separation below).

Zermelo's set theory (often denoted Z) is still an interesting object of study, but for mathematical purposes it is too weak, as was soon discovered. For example, it does not allow the construction of the union of an infinite family of sets in general.

---

[1]There is a real life version of the same paradox, about the "village barber, who shaves every villager who does not shave himself". Does the barber shave himself?

In 1922 the additional Axiom Scheme of Replacement was proposed by Abraham Fraenkel (1891–1965). The resulting system is called Zermelo–Fraenkel set theory, and denoted ZF. When the Axiom of Choice is added, we write ZFC.

The theory ZFC is formulated in the language $\{\in\}$, where $\in$ is a 2-place relation symbol expressing 'is an element of'. We only talk about sets and elementhood. We now list the axioms.

1. *Axiom of Extensionality*
   $$\forall x \forall y (\forall z (z \in x \leftrightarrow z \in y) \to x = y)$$
   Sets are equal if they have the same elements.
2. *Axiom of Pairing*
   $$\forall x \forall y \exists z \forall w (w \in z \leftrightarrow (w = x \lor w = y))$$
   For each x and y, $\{x, y\}$ is a set.
3. *Axiom Scheme of Separation*
   For every formula ϕ not containing the variable y, we have an axiom
   $$\forall x \exists y \forall w (w \in y \leftrightarrow (w \in x \land \phi))$$
   For each set x and property ϕ, $\{w \in x \mid \phi\}$ is a set.
4. *Axiom of Union*
   $$\forall x \exists y \forall w (w \in y \leftrightarrow \exists z (z \in x \land w \in z))$$
   For every set x, $\bigcup x$ (or $\bigcup_{z \in x} z$) is a set.
5. *Axiom of Power Set*
   $$\forall x \exists y \forall w (w \in y \leftrightarrow \forall z (z \in w \to z \in x))$$
   For every set x, $\mathcal{P}(x)$ is a set.
6. *Axiom of Infinity*
   Since $\exists x (x = x)$ is a valid sentence, there is a set. If x is a set then by Separation there is a set $\{w \in x \mid \perp\}$ which has no elements; and this set is unique by Extensionality. We denote this empty set by $\emptyset$. Also, for any set x we have a set $\{x\} = \{x, x\}$ by Pairing, and $x \cup \{x\} = \bigcup \{x, \{x\}\}$ using Pairing and Union. With these notations, the axiom of Infinity is now

   $$\exists x (\emptyset \in x \land \forall y (y \in x \to (y \cup \{y\}) \in x)).$$

   Suppose x is such a set as postulated by the axiom of Infinity. Then $\emptyset \in x$; let us write 0 for $\emptyset$. Then, also $0 \cup \{0\} \in x$; we write 1 for this set. Inductively, we write $n + 1$ for $n \cup \{n\}$. We can prove that n has exactly n elements, and therefore, that the set x "is infinite" in the sense that we can write down infinitely many elements of x (which are pairwise distinct).
7. *Axiom Scheme of Replacement*
   For any formula ϕ which does not contain the variable y:

   $$\forall a \exists b \forall c (\phi(a, c) \leftrightarrow c = b) \to$$
   $$\forall x \exists y \forall v (v \in x \to \exists u (u \in y \land \phi(v, u)))$$

The premiss expresses that ϕ defines an operation F on sets. The axiom says that for any such operation F and any set x, there is a set y which contains $\{F(v) \mid v \in x\}$ (it follows then by Separation that the latter is in fact a set).

8. *Axiom of Regularity*

$$\forall x (x \neq \emptyset \to \exists y (y \in x \wedge \forall z \neg (z \in y \wedge z \in x)))$$

Every nonempty set x has an element that is disjoint from x.

The regularity axiom (together with Pairing) implies that no set can be an element of itself, for if $x \in x$ then the set $\{x\}$ does not contain an element disjoint from itself. This has the following consequence:

There can be no "set of all sets" because such a set would be an element of itself. We see that the Russell paradox is resolved: the "set" $R = \{x \mid x \notin x\}$ would have to be the set of all sets! Therefore, R is not a set.

Remark 4.1.1 What does it mean to say that this theory is a "foundation for mathematics"? It means that all constructions from the basic set theory we developed in Chapter 1 can be defined in ZFC, and that all sets and functions used in mathematics can be regarded as elements of any model of ZFC. It is therefore possible to pretend that every mathematical theorem is a theorem about sets.

In principle, we could now translate every informal statement about sets in Chapter 1 into a formula of ZFC, and prove it from the ZFC-axioms by natural deduction trees. This is long and tedious but possible. Let us just stress these two points here:

1. The theory ZF, augmented by the Axiom of Choice where necessary, suffices to prove all the theorems and propositions of Chapter 1.
2. Now that we have formulated ZF(C) as a first-order theory, the question of whether or not a particular statement can be proved from it acquires a precise mathematical meaning.

Classes and Sets A *class* is a collection of all sets satisfying a given property. For us, a class is given by a formula $\phi(x)$ with one free variable x. Such a class is a set if $\exists y \forall x (\phi(x) \leftrightarrow x \in y)$ holds. Note that, by Separation, every subclass of a set is a set.

Using Pairing, we have for each set x and each set y the set

$$\{\{x\}, \{x, y\}\}$$

which we denote by (x, y) and call the *ordered pair* of x and y.

Exercise 130 Show:

(a) $(x, y) = (u, v) \leftrightarrow x = u \wedge y = v$.
(b) $x \times y = \{(u, v) \mid u \in x \wedge v \in y\}$ is a set. [Hint: use that for $u \in x$ and $v \in y$, the ordered pair (u, v) is a subset of $\mathcal{P}(x \cup y)$.]

The formulas $\exists x(x \in y \land \phi)$ and $\forall x(x \in y \to \phi)$ will be abbreviated to $\exists x \in y\phi$ and $\forall x \in y\phi$, respectively.

A *relation* from x to y is a subset of the set $x \times y$. Such a relation R is a *function* if $\forall u \in x \exists v \in y \forall w \in y((u, w) \in R \leftrightarrow v = w)$ holds.

Exercise 131 Show that for every pair of sets x, y there is a set y^x of all functions from x to y.

The *Axiom of Choice* can now be formulated:

$$\forall x \forall y \forall f \in y^x[\forall v \in y \exists u \in x((u, v) \in f) \to$$
$$\exists g \in x^y \forall v \in y \forall u \in x((v, u) \in g \to (u, v) \in f)].$$

A *poset* is an ordered pair (x, r) such that $r \in \mathcal{P}(x \times x)$ is a relation which partially orders x: i.e. $\forall u \in x((u, u) \in r)$, etcetera. Similarly, we can define the notions of a linear order and of a well-order.

Exercise 132 Carry this out.

Remark on Notation We have started to use a lot of symbols which are not part of the language $\{\in\}$: $\mathcal{P}(x)$, $\bigcup x$, $\{y \in x \mid \ldots\}$, $x \subseteq y$, etc. You should view these as abbreviations. Everything we express with these symbols can, equivalently, be said without them. For example, if $\phi(v)$ is a formula then the expression $\phi(\mathcal{P}(x))$ is short for:

$$\exists y[\forall v(v \in y \leftrightarrow \forall w(w \in v \to w \in x)) \land \phi(y)]$$

or equivalently

$$\forall y[\forall v(v \in y \leftrightarrow \forall w(w \in v \to w \in x)) \to \phi(y)].$$

Exercise 133 The list of axioms we have given is not the shortest possible. Show that the axiom of Pairing follows from Separation, Power Set and Replacement.

4.2 Ordinal Numbers and Cardinal Numbers

A set x is *transitive* if $\forall y \in x \forall u \in y(u \in x)$ holds.

Exercise 134 Prove that x is transitive iff $\forall y \in x(\mathcal{P}(y) \subset \mathcal{P}(x))$; and also that x is transitive iff $x \subset \mathcal{P}(x)$.

Examples. $\emptyset$ is transitive; $\{\emptyset\}$ too. $\{\{\emptyset\}\}$ is not transitive. If x and y are transitive, so is $x \cup y$, and if x is transitive, so is $x \cup \{x\}$.

A set x is an *ordinal number* (or just *ordinal*) if x is transitive and well-ordered by the relation $\in$. This means: x is an ordinal if the conditions

$$\forall y \in x \forall v \in y (v \in x)$$
$$\forall y \subseteq x (y \neq \emptyset \to \exists v \in y \forall w \in y (v \neq w \to v \in w))$$

hold.

Exercise 135 Check that these conditions indeed imply that $\in$ is a linear order on x, and that it is a well-order.

Normally, we use Greek lower-case characters in the first half of the alphabet: α, β, γ,... for ordinals.

Theorem 4.2.1

(a) $\emptyset$ *is an ordinal.*
(b) *If α is an ordinal then every $\beta \in \alpha$ is an ordinal.*
(c) *If α and β are ordinals then $\alpha \subseteq \beta \to \alpha = \beta \vee \alpha \in \beta$.*
(d) *If α and β are ordinals then $\alpha \subseteq \beta$ or $\beta \subseteq \alpha$.*

Proof. (a) is immediate and (b) is left as an exercise.

For (c), suppose $\alpha \subseteq \beta$ and $\alpha \neq \beta$. Let γ be the $\in$-least element of $\beta - \alpha$. Then $\gamma \subseteq \alpha$. On the other hand, if $x \in \alpha$ then $x = \gamma$ and $\gamma \in x$ are impossible by definition of γ and the fact that α is transitive. Therefore, since β is an ordinal, $x \in \gamma$ must hold. So $\alpha \subseteq \gamma$; hence $\alpha = \gamma$ and so $\alpha \in \beta$ as required.

Part (d) is proved by similar reasoning: suppose $\alpha \not\subseteq \beta$. Let γ be an $\in$-minimal element of $\alpha - \beta$. Then $\gamma \subseteq \beta$. If $\gamma = \beta$ then $\beta \in \alpha$ hence $\beta \subsetneq \alpha$ by transitivity of α. If $\gamma \subsetneq \beta$ then $\gamma \in \beta$ by (c), which contradicts the choice of γ. ∎

Exercise 136 Prove part (b) of Theorem 4.2.1. Prove also that if α and β are ordinals, then one of the following three possibilities must hold: $\alpha \in \beta$, $\alpha = \beta$, $\beta \in \alpha$.

Let Ord be the class of ordinal numbers. For ordinals α, β we write $\alpha < \beta$ for $\alpha \in \beta$. By Theorem 4.2.1, $<$ is a linear order on Ord. It is, actually, in a sense a well-order, as follows from the next theorem.

Theorem 4.2.2

(a) *Every nonempty subclass of Ord has a $<$-least element; in fact, if C is a nonempty class of ordinals, then $\bigcap C$ belongs to C.*
(b) *For every set x of ordinals, $\bigcup x$ is an ordinal, and it is the least ordinal α such that $\beta \leq \alpha$ for all $\beta \in x$.*
(c) *For every ordinal α, the set $\alpha \cup \{\alpha\}$ is an ordinal, and it is the least ordinal $> \alpha$. This set is denoted $\alpha + 1$.*

Proof. For (a), let C be defined by the formula $\phi(x)$ so we have $\forall x (\phi(x) \to x$ is an ordinal). Since C is nonempty, pick an x such that $\phi(x)$ holds. Then x is an ordinal and $x \cap C = \{y \in x \mid \phi(y)\}$ is a set, a subset of x. If $x \cap C = \emptyset$, then

x is the least element of C. Otherwise, since x is an ordinal, $x \cap C$ has an $\in$-least element in x. Then that element is the least element of C, as is left for you to check. We leave the remainder of the proof as an exercise. ∎

Exercise 137 Fill in the details of the proof above for part (a); prove parts (b) and (c) of Theorem 4.2.2.

Theorem 4.2.2 suggests that, in analogy with Theorem 1.4.6, there might also be a principle of "definition by recursion on Ord". This is in fact the case, but requires a little care in formulating.

Recall (from the introduction to the axiom of Replacement) that a formula $\phi(x, y)$ defines an operation on sets if

$$\forall x \exists y \forall z (\phi(x, z) \leftrightarrow y = z)$$

holds. We say that $\phi(x, y)$ defines an operation on ordinals if

$$\forall x (x \in \text{Ord} \rightarrow \exists y \forall z (\phi(x, z) \leftrightarrow y = z))$$

holds, where "$x \in \text{Ord}$" is the formula expressing that x is an ordinal.

Suppose $\phi(x, y)$ defines an operation on sets, which we call F. Then we use expressions like $\{F(x) \mid x \in y\}$ as shorthand; if ψ is a formula, the expression $\psi(\{F(x) \mid x \in y\})$ should be taken to mean

$$\exists z (\forall w (w \in z \leftrightarrow \exists x (x \in y \land \phi(x, w)))) \land \psi(z)).$$

Theorem 4.2.3 (Transfinite recursion on Ord) *For every operation F on sets there is a unique operation G on* Ord *such that for all ordinals α the following holds:*

$$G(\alpha) = F(\{G(\beta) \mid \beta \in \alpha\}).$$

Proof. Given an ordinal β, a set y and an element f of the set y^β of functions $\beta \rightarrow y$, we write $f[\beta]$ for the set $\{f(\gamma) \mid \gamma \in \beta\}$. Now define G by the following formula $\psi(\alpha, x)$:

$$\exists y \exists f \in y^\alpha [\forall \xi \in \alpha (f(\xi) = F(f[\xi])) \land x = F(f[\alpha])].$$

The proof that ψ defines an operation G on Ord with the stated property is left to you. ∎

Examples of Ordinals. $0 = \emptyset$, $1 = \{0\} = \{\emptyset\}$, $2 = \{\emptyset, \{\emptyset\}\} = \{0, 1\}$, $3 = \{0, 1, 2\}, \ldots$ are ordinals. Let x be a set as postulated by the axiom of Infinity, so $\emptyset \in x \land \forall y (y \in x \rightarrow y \cup \{y\} \in x)$. Let ω be the intersection of all subsets of x which contain $\emptyset$ and are closed under the operation $y \mapsto y \cup \{y\}$:

$$\omega = \{u \in x \mid \forall r \in \mathcal{P}(x)((\emptyset \in r \land \forall v (v \in r \rightarrow v \cup \{v\} \in r)) \rightarrow u \in r)\}.$$

Exercise 138

(a) Prove that ω is transitive.
(b) Prove that ω is an ordinal.

We think of the ordinal ω as the set of natural numbers:

$$\omega = \{0, 1, 2, \ldots\}.$$

By Theorem 4.2.2(c), we then also have the ordinals $\omega + 1$, $\omega + 2, \ldots$. One can show (using Replacement) that there is a set of ordinals $\{\omega + n \mid n \in \omega\}$ and hence (by Theorem 4.2.2(b)) an ordinal $\bigcup\{\omega + n \mid n \in \omega\}$, which we denote by $\omega + \omega$ or $\omega \cdot 2$.

Similarly we have $\omega \cdot 3$, $\omega \cdot 4, \ldots$ and $\bigcup\{\omega \cdot n \mid n \in \omega\}$, which we write as $\omega \cdot \omega = \omega^2$. Next, we get ω^3, $\omega^4, \ldots$ and $\bigcup\{\omega^n \mid n \in \omega\}$ which we write as ω^ω. Note that we have a certain ambiguity in notation here: ω^ω is *not* the set of all functions from ω to ω, but a *countable ordinal*.

We can continue: we have ω^{ω^ω} which is $\bigcup\{\omega^{\omega^n} \mid n \in \omega\}$, and so forth, leading to

$$\varepsilon_0 = \omega^{\omega^{\cdot^{\cdot^{\cdot}}}}.$$

All these ordinals are countable!

Exercise 139 (Addition and Multiplication of ordinals) By transfinite recursion (Theorem 4.2.3) we define operations of addition and multiplication on Ord, as follows:

$$\alpha + \beta = \begin{cases} \alpha & \text{if } \beta = 0 \\ \bigcup\{(\alpha + \gamma) + 1 \mid \gamma \in \beta\} & \text{otherwise} \end{cases}$$

and

$$\alpha \cdot \beta = \begin{cases} 0 & \text{if } \beta = 0 \\ \bigcup\{(\alpha \cdot \gamma) + \alpha \mid \gamma \in \beta\} & \text{otherwise.} \end{cases}$$

(a) Show that $\gamma < \beta$ implies $\alpha + \gamma < \alpha + \beta$, and (if $\alpha \neq 0$) $\alpha \cdot \gamma < \alpha \cdot \beta$.
(b) Show: $0 + \beta = \beta$ and $0 \cdot \beta = 0$.
(c) Show: $\alpha + (\beta + 1) = (\alpha + \beta) + 1$ and $\alpha \cdot (\beta + 1) = (\alpha \cdot \beta) + \alpha$.
(d) Show that for any nonempty set of ordinals x,

$$\alpha + \bigcup x = \bigcup\{\alpha + \beta \mid \beta \in x\}.$$

(e) Show that for any set of ordinals x,

$$\alpha \cdot \bigcup x = \bigcup\{\alpha \cdot \beta \mid \beta \in x\}.$$

(f) Show that $1 + \omega = \omega \neq \omega + 1$, and $2 \cdot \omega = \omega \neq \omega \cdot 2$.

(g) Show that $\alpha \cdot (\beta + \gamma) = \alpha \cdot \beta + \alpha \cdot \gamma$.

(h) Give a counterexample to the statement $(\alpha + \beta) \cdot \gamma = \alpha \cdot \gamma + \beta \cdot \gamma$.

Exercise 140 We can also define the operation of *exponentiation of ordinals*, as follows:

$$\alpha^\beta = \begin{cases} 1 & \text{if } \beta = 0 \\ \bigcup \{\alpha^\gamma \cdot \alpha \mid \gamma \in \beta\} & \text{otherwise.} \end{cases}$$

(a) Show that the notation ω^ω used before is actually an example of ordinal exponentiation.

(b) Show that $\alpha^{\beta + \gamma} = \alpha^\beta \cdot \alpha^\gamma$.

(c) Show that $(\alpha^\beta)^\gamma = \alpha^{\beta \cdot \gamma}$.

Theorem 4.2.4 *Every well-ordered set is isomorphic (as well-ordered set) to a unique ordinal number.*

Proof. Let $(X, <)$ be a well-ordered set. We use the principle of induction over X to show that for each $x \in X$ there is a unique ordinal $F(x)$ such that $\{y \in X \mid y < x\} \cong F(x)$. For successor elements $x + 1$, let $F(x + 1) = F(x) + 1 = F(x) \cup \{F(x)\}$. If l is a limit element, one proves that $\{F(x) \mid x < l\}$ is an ordinal which is isomorphic to $\{y \in X \mid y < l\}$. By similar reasoning, $\{F(x) \mid x \in X\}$ is an ordinal (it is a set by the Replacement scheme) which is isomorphic to $(X, <)$. ∎

Now recall Hartogs' Lemma, which states that for any set X there is a well-order $(W, <)$ such that W cannot be mapped injectively into X; by Theorem 4.2.4 there is an ordinal which cannot be mapped injectively into X, and by Theorem 4.2.2(a), there is a least such ordinal. Taking $X = \omega$, we see that there is a *least uncountable ordinal*, which we denote by ω_1.

The ordinals $0, 1, 2, \ldots, \omega$ and ω_1 are examples of *cardinal numbers*. A cardinal number is an ordinal κ such that for every $\alpha \in \kappa$, there is no bijection between α and κ.

If one assumes the Axiom of Choice, every set X can be well-ordered and is therefore in bijective correspondence with an ordinal; taking the least such ordinal, one associates to every set X a unique cardinal number κ such that there is a bijection between X and κ; we may write $|X|$ for κ. If we write 2^κ for $|\mathcal{P}(\kappa)|$ then the Continuum Hypothesis has a compact formulation:

$$2^\omega = \omega_1.$$

Without the Axiom of Choice one can still formulate the Continuum Hypothesis but one can no longer ascribe to every set a unique cardinal number as above.

There is a 1–1, surjective mapping from the class Ord of ordinals into the class of all infinite cardinal numbers, defined as follows: $\aleph_0$ (pronounced "aleph–zero") is ω. If $\aleph_\alpha$ is defined, $\aleph_{\alpha+1}$ is the least cardinal number greater than $\aleph_\alpha$. If λ is a limit ordinal (that is, an ordinal not of the form $\alpha + 1$), then $\aleph_\lambda = \bigcup\{\aleph_\beta \mid \beta < \lambda\}$.

Exercise 141 Show that $\aleph_\alpha$ is a cardinal for each α. Show also that for each infinite cardinal κ there is a unique ordinal α such that $\kappa = \aleph_\alpha$.

One can prove, without the Axiom of Choice, that $|\aleph_\alpha \times \aleph_\alpha| = \aleph_\alpha$ for each α. You should compare this to Proposition 1.3.9(iii).

4.2.1 The Cumulative Hierarchy

By Transfinite Induction on Ord, we define for every ordinal α a set V_α:

$$V_0 = \emptyset$$
$$V_{\alpha+1} = \mathcal{P}(V_\alpha)$$
$$V_\lambda = \bigcup_{\alpha \in \lambda} V_\alpha \text{ for a limit ordinal } \lambda.$$

Exercise 142 Prove the following facts:

(a) The set V_α is transitive, for every ordinal number α.
(b) For $\alpha \leq \beta$, we have $V_\alpha \subseteq V_\beta$.
(c) For $\alpha < \beta$, we have $V_\alpha \in V_\beta$.
(d) For every ordinal α, we have $\alpha \subseteq V_\alpha$.
(e) For every ordinal α, we have $\alpha = V_\alpha \cap \text{Ord}$.

Exercise 143 Prove, using the axioms of Regularity and Replacement, that every set is an element of some V_α.

Using Exercise 143 (and the fact that each V_α is transitive), we have that for every set x there is an α such that $x \subseteq V_\alpha$ and therefore there is a least such α. This ordinal is called the *rank* of x. Many statements about sets can be proved by induction on the rank of a set.

Exercise 144

(a) Suppose the rank of a set x is an infinite limit ordinal. Show that x is infinite.
(b) Suppose the rank of x is ω_1, the first uncountable ordinal. Show that x is uncountable.
(c) Prove that the converses of (a) and (b) do not hold.

Exercise 145 An ordinal number α is called *regular* if for no ordinal $\beta < \alpha$ there is a function $f : \beta \to \alpha$ with the property that $\forall y \in \alpha \exists x \in \beta (y \subseteq f(x))$.

(a) Show that every regular ordinal is a cardinal number.
(b) Show that ω_1 is regular.

4.3   The Real Numbers

The real numbers are constructed as follows. From ω, construct $\mathbb{Z}$ as the set of equivalence classes of $\omega \times \omega$ under the equivalence relation: $(n, m) \sim (k, l)$ iff $n + l = m + k$. Then there are well-defined operations of addition and multiplication on $\mathbb{Z}$. Define an equivalence relation on the set of those pairs (k, l) of elements of $\mathbb{Z}$ such that $l \neq 0$, by putting $(k, l) \sim (r, s)$ iff $ks = lr$. The set of equivalence classes is $\mathbb{Q}$, the set of rational numbers. $\mathbb{Q}$ is an *ordered field*, that is, a field with a linear order $<$ such that

(i) $r > s \rightarrow r + t > s + t$
(ii) $r > s > 0, t > 0 \rightarrow rt > st$

hold.

A *Dedekind cut* in $\mathbb{Q}$ (Richard Dedekind, 1831–1916) is a nonempty subset $A \subset \mathbb{Q}$ such that:

(i) $a \in A, a' < a \rightarrow a' \in A$
(ii) $\mathbb{Q} - A \neq \emptyset$
(iii) $\forall a \in A \exists b \in A(a < b)$

$\mathbb{R}$ is the set of Dedekind cuts in $\mathbb{Q}$, ordered by inclusion. $\mathbb{Q}$ is included in $\mathbb{R}$ via the embedding $\iota : q \mapsto \{r \in \mathbb{Q} \mid r < q\}$.

Exercise 146 Show that there are operations $+, \cdot$ on $\mathbb{R}$, making $\mathbb{R}$ into an ordered field which extends the ordered field $\mathbb{Q}$.

Suppose $\mathcal{A}$ is a set of elements of $\mathbb{R}$ which has an upper bound. Then $\bigcup \mathcal{A}$ is an element of $\mathbb{R}$: it is the least upper bound of $\mathcal{A}$. So $\mathbb{R}$ is complete. Moreover, $\mathbb{Q}$ is dense in $\mathbb{R}$: if $A, B \in \mathbb{R}$ and $A \subsetneq B$, there is a $q \in \mathbb{Q}$ such that $A \subsetneq \iota(q) \subsetneq B$. From this, it follows that $\mathbb{R}$ is a so-called *Archimedean ordered field*: that is, an ordered field such that for each a there is a natural number n such that $a < \iota(n)$. The following theorem, stated without proof (but the proof is not hard), characterizes the real numbers up to isomorphism.

Theorem 4.3.1 *There exists, up to order-isomorphism, exactly one complete Archimedean ordered field, the field of real numbers.*

We hope to have convinced you that everything we did in Chapter 1 can, at least in principle, be defined (and proved) within ZFC.

Appendix
Topics for Further Study

A.1 Incompleteness and Computability

The theory PA (see Example 2.5.4) might be seen as the theory of those truths that can be established by "finitary" (simple, constructive) methods. This gives it some philosophical significance. In studying PA, the class of "primitive recursive functions" is important. A function F of $n + 1$ variables is said to be defined *by primitive recursion* over functions G and H if for all numbers $x_1, \ldots, x_n, y$ we have

$$F(x_1, \ldots, x_n, 0) = G(x_1, \ldots, x_n)$$
$$F(x_1, \ldots, x_n, y + 1) = H(x_1, \ldots, x_n, F(x_1, \ldots, x_n, y), y).$$

We have basic functions: the constant zero function, the successor function and projections $(x, \ldots, x_n) \mapsto x_i$. A function is called *primitive recursive* if it can be obtained from the basic functions in a finite number of definitions by primitive recursion and substitution.

The relevance of primitive recursive functions for PA resides in the following two facts, both established by Gödel.

(i) Every primitive recursive function $F(x_1, \ldots, x_n)$ is *representable* in PA, that is: there is a formula $\phi_F(x_1, \ldots x_n)$ such that for each $n + 1$-tuple $m_1, \ldots, m_n, k$ of natural numbers, the sentence $\phi_F(\overline{m_1}, \ldots, \overline{m_n}, \overline{k})$ is provable in PA exactly when $F(m_1, \ldots, m_n) = k$. Here $\overline{k}$ stands for the PA-term $\underbrace{1 + \cdots + 1}_{k \text{ times}}$.

(ii) It is possible to encode terms, formulas and proof trees by natural numbers in such a way that all relevant information can be extracted by primitive recursive functions acting on the codes. We have a formula $\mathrm{Prf}(y, x)$ which says: x is the code of a PA-sentence ϕ and y is the code of a proof tree for ϕ.

© Springer Nature Switzerland AG 2018
I. Moerdijk, J. van Oosten, *Sets, Models and Proofs*, Springer Undergraduate Mathematics Series, https://doi.org/10.1007/978-3-319-92414-4

Alan Turing

Next, Gödel established that for every formula $\phi(x)$ in one free variable x there is a sentence ψ such that, if we write $\ulcorner\psi\urcorner$ for the code of ψ,

$$\text{PA} \vdash \psi \leftrightarrow \phi(\overline{\ulcorner\psi\urcorner}).$$

Now, apply this to the formula $\neg\exists y\,\text{Prf}(y, x)$. We obtain a sentence G (the famous "Gödel sentence") which satisfies

$$\text{PA} \vdash G \leftrightarrow \neg\exists y\,\text{Prf}(y, \overline{\ulcorner G\urcorner}).$$

In particular, the equivalence $G \leftrightarrow \neg\exists y\,\text{Prf}(y, \overline{\ulcorner G\urcorner})$ is true in the standard model $\mathbb{N}$; so G is true precisely when there is no proof for G. It follows that there can be no proof for G (otherwise, G would be false, and PA would prove false things), hence G is true, and PA is incomplete as a theory. This is Gödel's *First Incompleteness Theorem*, published in [23].

Gödel's *Second Incompleteness Theorem* is the statement that the Gödel sentence G, as given above, is equivalent (in PA) to the sentence $\neg\exists y\,\text{Prf}(y, \overline{\ulcorner\bot\urcorner})$, a sentence which expresses that the theory PA is formally consistent. This result, in combination with the fact that the sentence G has no proof, is often paraphrased by saying "PA does not prove its own consistency". Good technical treatments of Gödel's Theorems are in [4] and [27]. An amusing book, dealing with the sense and nonsense that have been made out of Gödel's theorems in philosophy and the arts, is [20].

Tarski noted that Gödel's proof admitted the following corollary: there cannot be a formula Tr(x) in one free variable which defines those sentences which are true in $\mathbb{N}$ (that is: for every sentence ϕ we would have the equivalence $\phi \leftrightarrow \text{Tr}(\overline{\ulcorner\phi\urcorner})$), because by Gödel's observation there would be a sentence ψ such that $\psi \leftrightarrow \neg\text{Tr}(\overline{\ulcorner\psi\urcorner})$ would be a theorem of PA. One says: "truth is not definable" (and it is a running joke that Tarski both gave a definition of truth and proved the undefinability of truth).

In the wake of Gödel's results, people started thinking about what it means to calculate something according to an algorithm. Definitions were proposed almost simultaneously by several authors, but the most imaginative and convincing paper was written by *Alan Turing*: [78]. [Alan Mathison Turing (1912–1954) was a brilliant mathematician who also gained fame during the second world war due to his work on deciphering the German "Enigma" coding machine. His life ended tragically. For a biography, see [35].] Turing envisages a "computer" (which is, for him, a person who calculates) and argues that such an individual must be in one of only finitely many "states of mind". He scans a notebook which has places for symbols from a finite alphabet, and at each place, his state of mind together with the symbol he reads determine his action (erase the symbol or replace it by another, and move to the next place or the previous one) and the next state of mind he is in. One state of mind has particular importance: the *halting state*. The computer stops, and the contents of the notebook at that point are the result of the computation. Later, this informal description led to the formal definition of a *Turing machine*, a finite automaton. Turing argues cogently for the *Church–Turing Thesis*, which asserts that anything that can be calculated algorithmically at all, can be calculated by a Turing machine. A function $f : \mathbb{N}^k \to \mathbb{N}$ is called *computable* if there exists one Turing machine which computes $f(n_1, \ldots, n_k)$ for every input $n_1, \ldots, n_k$.

An equivalent definition is the following. A function $f : \mathbb{N}^k \to \mathbb{N}$ is said to be defined from $g : \mathbb{N}^{k+1} \to \mathbb{N}$ by *minimalization* (or *μ-recursion*) if for all inputs $n_1, \ldots, n_k$, we have that $f(n_1, \ldots, n_k)$ is the least number m such that $g(n_1, \ldots, n_k, m) = 0$. A function f is called *general recursive* if it can be defined by minimalization and substitution from primitive recursive functions. A set X of k-tuples of natural numbers is called *decidable* if its characteristic function is general recursive. Turing gave the following example of a non-decidable set: the *Halting Set*. This is the set of those pairs (a, b) of natural numbers which satisfy: the b-th Turing machine with input a reaches the halting state. There is no algorithm to decide, for an arbitrary algorithm and arbitrary input, whether the computation ever ends!

Of course, mathematicians have known and used algorithms since antiquity (think of long division, Euclid's algorithm for the greatest common divisor, bisection of an angle, etcetera); but the formal definition can be used to explore the boundaries of the subject, and to prove that for particular problems, *there can be no algorithm to solve them*. As an example, consider the Tenth Problem posed by Hilbert in 1900 (as quoted in the Introduction of this book):

10. Find an algorithm to determine whether a polynomial equation with integer coefficients has a solution in the integers.

After decades of effort and much preparatory work by M. Davis, H. Putnam and J. Robinson, it was finally the young Russian Yuri Matiyasevich (he was 23 at the time!) who proved, in 1970, the following theorem (we give a simplified form):

Theorem A.1.1 *There is a natural number $k \geq 1$ and a polynomial*

$$H(X_1, X_2, Y_1, \ldots, Y_k)$$

in $k + 2$ variables and with integer coefficients which has the following property: for a pair of natural numbers (a, b), (a, b) is in Turing's Halting Set if and only if the polynomial equation $H(a, b, Y_1, \ldots, Y_k) = 0$ has a solution in the integers.

From this theorem, in combination with the fact that the Halting Set is undecidable, we see that an algorithm such as asked for in Hilbert's Tenth Problem cannot exist. For an exposition of Theorem A.1.1, see the book [53] or (better) the lecture notes [52].

Matiyasevich's result has given rise to a whole branch of research, which started by proving or disproving Theorem A.1.1 when $\mathbb{Z}$ is replaced by the ring of integers of a number field. This is a happy marriage of the theory of computable functions and algebraic number theory. For an overview of recent results, see [61].

Another offshoot of Gödel's and Turing's work concerns the *Entscheidungsproblem* (Decision Problem) posed by Hilbert: determine whether an arbitrary formula of first-order logic is valid. *Church's Theorem* states that this cannot be done by an algorithm.

The theory of recursive functions has branched in several directions. There is the theory of "relative computability" (what is algorithmically computable if a – non-computable – function is "given to us by an oracle"?); there is the theory of "higher-type computability" (what does it mean to say that a *functional* – i.e., a function acting on functions – is computable). One of the most authoritative textbooks is still [65]. Other good books are [56, 57, 72].

The incompleteness of PA has triggered the study of nonstandard models of PA. This is a rich area of research, in which Computability Theory, Number Theory and Model Theory come together in a wonderful blend. For an introduction in this field, see [40].

A.2   Proof Theory

Let us recall Hilbert's Programme (HP) from the Introduction to this book. Hilbert distinguished an *actual* mathematical world of finite (or finite-dimensional, combinatory) things, and an *ideal* world of infinitary abstractions. He proposed constructing a logical system S capable of describing the actual world, which includes proofs about the ideal world.

With the knowledge we have gained in this course we can now formulate a mathematical implementation of HP as follows: for the system S we take the theory

Gerhard Gentzen

PA. By Chapter 4 we can assume that the infinitary world is adequately represented in the system ZF (or ZFC). In this interpretation we have the following mathematical reformulation of HP:

> The *weak version* states that PA proves the consistency of ZF(C).
> The *strong version* states that ZF(C) is *conservative over* PA, and hence that PA is complete.[1]

In Section A.1 we saw that Gödel's First Incompleteness Theorem says that PA is *not* complete; therefore this theorem refutes the strong version of HP. Moreover, PA can be seen as a subsystem of ZF(C), so the consistency of ZF(C) implies that of PA. This means that Gödel's Second Incompleteness Theorem refutes the weak version of HP. It would thus appear that Hilbert's Programme was sunk by Gödel's results and this is indeed the common view among logicians, although [16] disagrees. For a comprehensive historical and philosophical exposition of HP, see [69]; see also [44, 71].

---

[1] Suppose that the formula ϕ represents an arithmetical truth. If the truth of ϕ can be established by ordinary mathematical reasoning, then ϕ is a theorem of ZF(C); by the conservativity, ϕ must be provable in PA.

All this does not mean that Hilbert's idea of *Proof Theory*, the combinatorial study of formal proofs, was an idle fantasy.

The first substantial advances were made by Gerhard Gentzen. First he showed that any formal proof can be brought into so-called *Normal Form*. Gentzen's system is different from our natural deduction trees in Chapter 3, so we will not go into details of what this normal form looks like; we just state two facts which can be seen as straightforward consequences of the Normal Form Theorem (or "Cut Elimination").

Theorem A.2.1 (Herbrand's Theorem)   *Let L be a language with at least one constant and let T be an L-theory consisting of sentences of the form $\forall x_1 \cdots \forall x_k \phi$, with ϕ quantifier-free. Suppose that*

$$T \vdash \forall x_1 \cdots \forall x_n \exists y_1 \cdots \exists y_m \psi(\vec{x}, \vec{y})$$

with ψ quantifier-free. Then there is a double sequence $(t_{ij} \mid 1 \le i \le r, 1 \le j \le m)$ of L-terms such that the formula

$$\bigvee_{i=1}^{r} \psi(\vec{x}, t_{i1}, \dots, t_{in})$$

is a consequence of T.

[Jacques Herbrand (1908–1931), who just lived for 23 years, was a mathematical prodigy. He had results in several mathematical fields: the Herbrand quotient in homological algebra, the Herbrand–Ribet theorem in number theory. He was an avid mountaineer and died by a fall in the Massif des Écrins.]

Theorem A.2.2 (Craig's Interpolation Theorem)   *Given two languages L and L', let ϕ be an L-sentence and ψ an L'-sentence. If $\vdash \phi \to \psi$ then there exists an $(L \cap L')$-sentence χ such that $\vdash \phi \to \chi$ and $\vdash \chi \to \psi$.*

[William Craig, 1918–2016] It should be said that this particular result can also be proved by model-theoretical means (although this was not how the original proof went).

Gentzen did important work in the proof theory of PA as well. Since the consistency of PA cannot be proved from the PA-axioms alone, any proof of it must have some "infinitary content". Gentzen set out to isolate and make explicit this infinitary part. Consider a primitive recursive binary relation $\prec$ (that is, a relation whose characteristic function is primitive recursive) on $\mathbb{N}$ such that $(\mathbb{N}, \prec)$ is a well-order. Then $(\mathbb{N}, \prec)$ is isomorphic to a unique ordinal number α; say that the relation $\prec$ *represents* α. Moreover, the relation $\prec$ is definable in PA by a formula $\phi(x, y)$. We say that "PA proves transfinite induction along α" (PA $\vdash$ TI(α)) if there exists a primitive recursive relation $\prec$ which represents α and a formula ϕ which defines $\prec$ such that for every formula $\psi(x)$ we have:

$$\text{PA} \vdash [\forall x (\forall y (\phi(y, x) \to \psi(y)) \to \psi(x))] \to \forall x \psi(x).$$

Gentzen was able to prove the following statements about PA and the ordinal number $\varepsilon_0 = \omega^{\omega^{\cdot^{\cdot}}}$:

Theorem A.2.3 (Gentzen) *Let* Con(PA) *be the statement which expresses the consistency of* PA. *Then the following hold:*

(i) *For any countable ordinal* α *we have* PA $\vdash$ TI(α) *if and only if* $\alpha < \varepsilon_0$.
(ii) PA $+$ TI(ε_0) $\vdash$ Con(PA).

Item (i) of the theorem above is often expressed by saying that ε_0 is the *proof-theoretic ordinal of* PA. For many arithmetical theories (subsystems of PA or extensions of PA) such an "ordinal analysis" has been carried out, although there are also systems which have defied every attempt. For good introductions to the field of ordinal analysis, see [60, 63].

We conclude this section by pointing out two modern developments where genuine mathematical applications arise out of the study of formal proofs: *Proof Mining* and *Formal Proof Verification*. In both cases, one studies completely formalized mathematical proofs (proofs which are written out in natural deduction trees or some other format).

In Proof Mining, such proofs undergo a judiciously chosen transformation (a so-called *proof interpretation*) after which it is often possible to extract extra information from the formal proof. The extra information can be a previously unknown algorithm, or sharper bounds for approximation theorems. The applications are mostly in Analysis or Functional Analysis. This field, initially proposed by Georg Kreisel (1923–2015), was brought to fruition by Ulrich Kohlenbach; the book [43] is an excellent exposition.

Formal Proof Verification exploits the (already mentioned) finitary character of proof systems to let the computer assist in constructing and validating completely formalized proofs. The program which checks that a given formal proof is correct is small (since a proof system consists of a set of proof rules that can very often be displayed on two A4 pages) and can therefore be verified "by hand". Therefore, if we have constructed a formal proof and have it checked, we can be certain that the result is correct, even if the original mathematical proof is so enormous that checking by hand is too time-consuming and error-prone to lead to anything trustworthy. For example, a proof of the Four-Colour Theorem was completely formalized and verified in 2004. Formal Proof Verification is now almost a mathematical topic in its own right, and it is practised in many places. For good, intuitive introductions, see the papers by Tom Hales and Freek Wiedijk: [28, 29, 84]. To quote Hales: *It has been necessary to [. . .] retool the foundations of mathematics for practical efficiency, while preserving its reliability and austere beauty.*

Two general introductions to Proof Theory (beyond ordinal analysis) are [68, 77]. Especially recommended are Buss' chapters [6, 7] in the Handbook of Proof Theory.

A.3   Model Theory

In Chapter 2 we have discussed models of first-order theories. The chapter ended
with a few glimpses into the subject of model theory, a rapidly developing subfield
of logic, with many connections to algebra and algebraic geometry. We discussed
elimination of quantifiers.

One example of quantifier elimination that we skipped, but which is also very
important, is the theory of *real closed fields*: the theory says that we have an ordered
ring which is a field, that $x < y$ if and only if $y - x$ is a square and that every monic
polynomial of odd degree has a root – in short, the theory of the ordered field $\mathbb{R}$.
The fact that this theory satisfies the condition of Lemma 2.7.4 is essentially the
same as what is known as the *Tarski–Seidenberg Principle*, which is the cornerstone
of the field of *Real Algebraic Geometry* (see [3]). One of the early pioneers of
Model Theory, Abraham Robinson, used this quantifier-elimination result to reprove
Hilbert's 17th Problem (see the Introduction). An earlier proof of the same result had
been given by Emil Artin (1898–1962) in 1927 [1].

We also discussed elementary embeddings between models, and we proved that
any two countable models of the theory of dense linear orders without end points are
isomorphic. The proof we presented used a back-and-forth argument relating finite

Abraham Robinson

subsets of two different linear orders, a first example of the much more general (Fraïssé) method of constructing isomorphisms between models. In Chapter 2 the formulation was that the theory of dense linear orders is "ω-categorical".

A famous theorem of Michael D. Morley (1930–) states that if a countable theory is categorical in some *un*countable cardinal, then it is categorical in *every* uncountable cardinal. All these topics, including Morley's theorem, form just a small part of "classical" model theory. The book *Saturated Model Theory* by G. Sacks [67] is a classic, and provides a concise introduction. Among the more comprehensive introductions to classical model theory, we recommend the book by Tent and Ziegler [76].

One of the central notions in model theory that we did not cover is that of a "type". If M is a model of a first-order L-theory and A is a subset of M, consider sets of formulas in the free variables $x_1, \ldots, x_n$, in the language L_A. Such a set may be *realized* in the model, meaning that there are elements $b_1, \ldots, b_n$ in M for which these formulas are all true. An n-type with parameters from A is a maximally consistent such set (consistent with the theory). Such a type may or may not be realized. The *saturated* models to which the title of Sacks' book refers are those models which realize all their types. But the dual notion of a model which only realizes the types it has to (such a model is called *atomic*) is equally important, and existence theorems for such models go under the name of the Omitting Types Theorem.

Morley's theorem leads to the question of understanding the countable models of a countable theory and the structure between these, and to the question to what extent these results hold for uncountable theories. The subject of *Stability Theory*, now one of the main areas within model theory, developed out of these questions. Based on the early work of Morley, this subject really came off the ground through the work of S. Shelah, who proved an analogue of Morley's result for uncountable theories and introduced several central notions, such as that of a *stable theory* – a theory for which the number of types with parameters in A is controlled by the size of A. For any L-theory T one can also consider n-types of T: maximal sets of L-formulas $\phi(x_1, \ldots, x_n)$ which are consistent with T. An example of how the number of types interacts with other model-theoretic structure is the *Ryll-Nardzewski Theorem*, which says that a countable theory is ω-categorical if and only if it has only finitely many n-types, for each n.

Good first introductions to Stability Theory are Pillay's notes [58] and the Tent–Ziegler book already mentioned.

For a structure M of a particular language L, the collection of definable subsets of M^n for various n has a lot of structure. For example, it is closed under the boolean operations and under taking the image of a projection $M^n \to M^{n-1}$. One can study families of subsets of Euclidean spaces having such a structure, leading to the theory of o-minimal structures, which can be viewed as one way of doing topology of these Euclidean spaces in a more controlled way, thus fitting into Grothendieck's program of *topologie modérée* [26]. The book [81] by Van den Dries is a very accessible introduction to this area. The use of such first-order definable sets is also central in the approach to *Motivic Integration* via Model Theory, developed by Denef and Loeser, see e.g. [80]. Related ideas lie at the origins of the theory of Zariski geometries developed by Hrushovski and Zilber. Here one starts with a collection of subsets of powers of an abstractly given set X satisfying closure conditions similar to those of the definable sets, and tries to find the underlying field for which these sets are the closed sets in a Zariski topology in associated algebraic varieties. We refer the interested reader to the book [87] of Zilber. The work of Hrushovski and Zilber forms part of an area called *Geometric Stability Theory*, where ideas from Model Theory and Algebraic Geometry naturally blend. We refer the interested reader to [5] for an accessible survey, and to [59] for a systematic exposition. A hallmark result in this area is Hrushovski's model-theoretic proof of the Mordell–Lang Conjecture in Algebraic Geometry; an exposition of this result can be found in [5].

A.4 Set Theory

In our first chapter we discussed several aspects of what is often termed "naïve" set theory, the theory of sets as used by working mathematicians on a day to day basis. Naïve set theory includes the study and use of various forms of the Axiom of Choice, but lacks a precise decision on the axiomatisation of the theory itself. The most common way to be more explicit about this is to adopt the axioms of ZFC, Zermelo–Fraenkel Set Theory with the Axiom of Choice, as presented in Chapter 4. Once one has agreed on an explicit axiomatization as a first-order theory it becomes possible to study the models (in the sense of Chapter 2) of this theory. In reasoning about these models one might use naïve set theory as one's ambient framework (the "meta-theory"), or one might prefer some more formal meta-theory such as ZFC itself. In the latter case, this leads to the study of models of ZFC inside ZFC, or of models of ZFC which are themselves elements of another model of ZFC. Different methods of constructing such models have led to the famous early independence and consistency results in set theory. One of the first examples is Gödel's method of constructing a very "thin" model inside any model of ZF. This thin model, called the universe L of *constructible sets*, always satisfies the Axiom of Choice and also the Continuum Hypothesis, which shows the consistency of these axioms (relative to ZF). A good book about the constructible sets is [17].

In the opposite direction, Cohen's method of forcing constructs a "wide" model of ZFC having many subsets of the natural numbers, in which the Continuum Hypothesis fails. Another way to think of Cohen's model is as a universe of sets parametrized by a large complete Boolean algebra. This parametrisation then provides enough room for there to be a set whose cardinality lies strictly in between the cardinality of the natural numbers and that of the real numbers. Yet another method to construct models is to study sets parametrized by a Boolean algebra on which a group acts. The sets given by parametrisations which are invariant under the group action form a model of ZF, but the choice functions provided by the ambient theory ZFC need not be invariant under the group action, and this leads to models

Paul Cohen

in which the Axiom of Choice is false. For a concise introduction to this kind of "model theory" of ZF(C), see [2].

These methods of constructing models can be iterated and combined in various ways, leading to many more consistency and independence results. Many of these methods of constructing models of set theory also have a more geometric, sheaf-theoretic interpretation, see [48].

The method of forcing rests on finding suitable "generic" filters in partially ordered sets. An axiom ensuring the existence of suitable generic sets is Martin's Axiom: it states that in a partially ordered set satisfying the *countable chain condition*, one can always find a filter intersecting any family of dense subsets as long as the cardinality of that family is strictly less than 2^ω, the cardinality of the set of real numbers. The notion of density here is the usual topological one for the Alexandrov topology on the partially ordered set; this is the topology where the open sets are the downward-closed ones. For this topology the countable chain condition says that any family of pairwise disjoint open subsets is (at most) countable. An equivalent formulation, more familiar to topologists, states that a compact Hausdorff space satisfying the countable chain condition cannot be the union of less than 2^ω many nowhere dense subsets. This is reminiscent of the Baire Category Theorem, which implies that in a separable compact Hausdorff space, the union of countably many nowhere dense open sets is still nowhere dense ([55], Theorem 48.2). In particular, Martin's Axiom is a consequence of the Continuum Hypothesis. But it is also consistent with the Axiom of Choice and the negation of the Continuum Hypothesis, so that Martin's Axiom can refer to larger than countable families of dense subsets. Martin's Axiom is also relevant in the study of non-principal ultrafilters of subsets of the natural numbers, i.e. of points of the Čech–Stone compactification of the natural numbers (for a survey on this space, see [82]).

The Axiom of Choice enables one to prove the existence of "strange" mathematical objects such as non-measurable subsets of the real line. One may wonder whether such mathematical objects "should" really be there, in particular since they cannot be explicitly be defined in ZFC [73]. An axiom which avoids the existence of such objects is the Axiom of Determinacy. This axiom plays an important role in the field of Descriptive Set Theory which studies "hierarchies" of definable subsets of Baire space or more general complete separable metric spaces. The most authoritative introduction to this field is [41].

The existence of so many different models of extensions of Zermelo-Fraenkel set theory all validating different properties of infinite cardinal numbers is somewhat at odds with the idea that naïve set theory forms a commonly agreed ("intersubjective", philosophers might say) basis for mathematics. This leads to the search for an extension of ZF, or of ZFC if one accepts the axiom of choice, given by axioms which can be justified on the basis of our mathematical intuition concerning the real line. Partly motivated by this, Set Theory has come up with a wealth of "large cardinal axioms" which assert the existence of infinite cardinals with specific properties: inaccessible cardinals, measurable cardinals, Woodin cardinals and many more. One can then investigate the effect that extensions of ZFC by axioms stating the existence of such "types of infinity" have on the consistency and proof theoretic strength of each other, and on the construction and properties of particular subsets of the real line. The literature on these large cardinals is vast and growing steadily. The reader might consult [39] for a first impression.

A.5 Alternative Logics

In this book we have restricted ourselves to languages of first-order logic, their models and their proof theory. We have seen that many mathematically interesting theories can be formulated in a first-order language, such as the theories of groups or rings, the theory Peano Arithmetic for the natural numbers, and Zermelo–Fraenkel set theory. We have only considered what are called "single-sorted" languages and their theories. A minor linguistic variation is that of many-sorted languages. Such a language has a set of "sorts" or "types" S_i (indexed by $i \in I$), infinitely many variables of each sort, and function symbols having a specified finite sequence of sorts as their domain and a specified sort as their codomain, and similarly for relation symbols. It is sometimes more natural to use such a multi-sorted language to formulate a mathematical theory: for example, the theory of graphs, or the theory of vector spaces. On the other hand, one can always reformulate a multi-sorted first-order theory as a single-sorted one, using unary relation symbols for the sorts.

Bertrand Russell

Another variation of first-order logic is where one allows infinite strings of quantifiers, and/or infinite disjunctions or conjunctions, as for example in the formula

$$\exists x_0 \exists x_1 \cdots \bigwedge_{i=0}^{\infty} x_{i+1} < x_i,$$

which makes it possible to define well-orders (we have seen in Exercise 81 that by the Compactness Theorem this is not possible in ordinary first-order logic). A recent source for more on such infinitary languages and their model theory is [50].

On the other hand, there are many mathematical subjects which do not immediately allow a first-order formulation. For example, in the description of a topological space one uses quantification over points of the space as well as over (open and closed) subsets of the space. Similarly, in Peano arithmetic one cannot formulate induction over arbitrary subsets of the natural numbers, but only over definable ones. These are examples where one can effectively use second-order languages, which do allow quantification over elements as well as over subsets (of the natural numbers, in the latter case). Motivated by Gödel's Incompleteness Theorems and questions about proof-theoretic strength, this "second-order Peano arithmetic" has been studied in detail, see e.g. [70].

A different approach to quantifying over subsets is to work in a many-sorted language and specify for each type X another type $P(X)$ and a binary membership relation symbol $\in_X$ (taking arguments of type X and $P(X)$), while introducing a so-called *Comprehension Axiom* to express that $P(X)$ acts as the power set of X:

$$\exists A{:}P(X)\forall x{:}X(x \in_X A \leftrightarrow \phi(x))$$

for every formula $\phi(x)$. Of course this can be iterated: one may have types $P(P(X))$, etcetera, leading to *higher-order logic*. Logics of this kind are closely related to logical interpretations of topos theory, see [46].

One can do something similar with regards to quantification over functions, and introduce for any two types X and Y a function type Y^X together with a function symbol $\mathrm{ev}_{Y,X} : Y^X \times X \to Y$ for the evaluation of functions at arguments in X, accompanied by a function comprehension axiom of the form

$$\forall x{:}X \exists! y{:}Y \phi(x, y) \;\to\; \exists f{:}Y^X \forall x{:}X \phi(x, \mathrm{ev}_{Y,X}(f, x)),$$

where the quantifier $\exists! y$ means "there exists exactly one y".

This formalism is closely related to theories in the so-called *typed lambda calculus*.

An example theory of the latter kind is the theory PA^ω ("Peano Arithmetic in all finite types") of the natural numbers, functions from numbers to numbers, etcetera. It has a type N for the natural numbers, and for any two types X and Y a function type Y^X, so in particular a type N^N for number theoretic functions. Among the models are the topological models (interpreting N^N as Baire space and $N^{(N^N)}$ as the space of continuous functions $N^N \to N$), and computable models interpreting N^N as computable functions. Theories of this kind have been studied in the references on Proof Theory given above.

The examples above of higher-order theories or type theories are really multi-sorted first-order theories in disguise, where one adds comprehension axioms to impose a relation between the types. A completely different approach to type theory is the one usually referred to as Martin-Löf Type Theory. Here the language has no formulas or sentences, but only types and terms (possibly depending on free variables) of specified types. The idea is that the types also express propositions, and each term of a given type witnesses that the type is non-empty: one can think of such a term as a proof of the proposition expressed by the type. Types can depend on variables ranging over another type, and quantification is formulated by operations on types. For example, if $Y(x)$ is a family of types depending on a variable x of type X, then the product type

$$\prod_{x \in X} Y(x)$$

plays the role of universal quantification over X. This formalism, sometimes referred to as "propositions-as-types", is very close to functional programming languages such as Agda. Theories of this kind have interesting topological models, where one interprets proofs of an equality between terms as homotopies between functions. We refer the reader to [79] for an exposition of Martin-Löf Type Theory and these "homotopical models".

Photo Credits

We are grateful to the following institutions for permissions to include photographs: the Mathematisches Forschungsinstitut Oberwolfach, Martin-Luther-Universität Halle-Wittenberg, the Institute of Advanced Study, Albert-Ludwigs-Universität Freiburg, King's College Cambridge (UK), Stanford University and McMaster University. Individual credits:

p.viii Photographer unknown. Source: Konrad Jacobs and the Mathematisches Forschungsinstitut Oberwolfach.

p.2 Photographer unknown. Source: Universitätsarchiv Halle.

p.38 Photographer: George M. Bergman. Source: Mathematisches Forschungsinstitut Oberwolfach.

p.82 Photographer unknown (gift of Judith Sachs and Gabrielle Forrest). Source: the Shelby White and Leon Levy Archives Center, Institute for Advanced Study, Princeton, N.J.

p.104 Photographer unknown. Source: Universitätsarchiv Freiburg C 129/338.

p.116 Photographer unknown. Source: the Archive Centre at King's College, University of Cambridge.

p.119 Photographer unknown. Source: Eckart Menzler-Trott and the Mathematisches Forschungsinstitut Oberwolfach.

p.122 Photographer: Konrad Jacobs. Source: Mathematisches Forschungsinstitut Oberwolfach.

p.125 Courtesy of the Cohen Family.

p.127 Photographer unknown. Source: William Ready Division of Archives and Research Collections, McMaster University Libraries, Canada.

© Springer Nature Switzerland AG 2018
I. Moerdijk, J. van Oosten, *Sets, Models and Proofs*, Springer Undergraduate Mathematics Series, https://doi.org/10.1007/978-3-319-92414-4

Bibliography

1. Artin, E.: Über die Zerlegung definiter Funktionen in Quadrate. Abh. Math. Sem. Univ. Hamburg **5**, 85–99 (1927)
2. Bell, J.L.: Set Theory – Boolean-Valued Models and Independence Proofs. Oxford Logic Guides, vol. 47. Clarendon Press, Oxford (2005)
3. Bochnak, J., Coste, M., Roy, M.-F.: Real Algebraic Geometry. Ergebnisse der Mathematik und ihre Grenzgebiete, vol. 36. Springer, New York (1998)
4. Boolos, G., Burgess, J.P., Jeffrey, R.C.: Computability and Logic, 5th edn. Cambridge University Press, Cambridge (2007)
5. Bouscaren, E. (ed.): Model Theory and Algebraic Geometry. LNM, vol. 1696. Springer, Berlin/Heidelberg (1998)
6. Buss, S.R.: An introduction to proof theory. In: Buss, S.R. (ed.) Handbook of Proof Theory, Chapter I. Elsevier, New York (1998)
7. Buss, S.R.: First-order proof theory of arithmetic. In: Buss, S.R. (ed.) Handbook of Proof Theory, Chapter II. Elsevier, New York (1998)
8. Cantor, G.: Über eine Eigenschaft des Inbegriffes aller reellen algebraischen Zahlen. Journal für die reine und angewandte Mathematik **77**, 258–262 (1874)
9. Cohen, P.J.: The independence of the continuum hypothesis. Proc. Nat. Acad. Sci. U.S.A. **50**, 1143–1148 (1963)
10. Cohen, P.J.: Independence results in set theory. In: Addison, J.W., Henkin, L., Traski, A. (eds.) Theory of Models, pp. 39–54. North-Holland, Amsterdam (1965)
11. Cohn, P.M.: Basic Algebra. Springer, London (2003)
12. Dauben, J.W.: Georg Cantor. Princeton University Press, Princeton (1990)
13. Davis, M. (ed.): The Undecidable – Basic Papers on Undecidable Propositions, Unsolvable Problems and Computable Functions. Dover, Mineola (2004) Reprint of 1965 edition by Raven Press Books
14. Dawson, J.W.: Logical Dilemmas – The Life and Work of Kurt Gödel. Peters, Wellesley (1997)
15. de Bruijn, N.G., Erdős, P.: A colour problem for infinite graphs and a problem in the theory of relations. Indagationes Mathematicae **13**, 369–373 (1951)
16. Detlefsen, M.: On an alleged refutation of Hilbert's program using Gödel's first incompleteness theorem. J. Philos. Log. **19**(4), 343–377 (1990)
17. Devlin, K.J.: Constructibility. Perspectives in Mathematical Logic. Springer, Berlin/Heidelberg (1984)
18. Ewald, W.B. (ed.): From Kant to Hilbert: A Source Book in the Foundations of Mathematics. Oxford Uinversity Press, Oxford (2000)

© Springer Nature Switzerland AG 2018
I. Moerdijk, J. van Oosten, *Sets, Models and Proofs*, Springer Undergraduate Mathematics Series, https://doi.org/10.1007/978-3-319-92414-4

19. Feferman, S.: Some applications of the notions of forcing and generic sets. Fundamenta Mathematicae **56**, 325–345 (1964)
20. Franzén, T.: Gödel's Theorem – An Incomplete Guide to Its Use and Abuse. AK Peters, Wellesley (2005)
21. Gentzen, G.: Untersuchungen über das logische Schließen. Mathematische Zeitschrift **39**, 176–210 (1934)
22. Gödel, K.: Die Vollständigkeit der Axiome des logischen Funktionenkalküls. Monatshefte für Mathematik und Physik **37**, 349–360 (1930)
23. Gödel, K.: Über formal unentscheidbare Sätze der Principia Mathematica und verwandter Systeme I. Monatshefte für Mathematik und Physik **38**, 173–198 (1931); English translation in [13]
24. Gödel, K.: The Consistency of the Continuum Hypothesis. Annals of Mathematics Studies, vol. 3. Princeton University Press, Princeton (1940)
25. Goldblatt, R.: Lectures on the Hyperreals – An Introduction to Nonstandard Analysis. Graduate Texts in Mathematics, vol. 188. Springer, New York (1998)
26. Grothendieck, A.: Esquisse d'un Programme (1984). Research Proposal (unpublished)
27. Hájek, P., Pudlák, P.: Metamathematics of First-Order Arithmetic. Perspectives in Mathematical Logic. Springer, Berlin/New York (1993); Second printing (1998)
28. Hales, T.C.: Formal proof. Not. AMS **55**(11), 1370–1380 (2008)
29. Hales, T.C.: Developments in formal proofs. Séminaire Bourbaki **2013–2014**(1086), 1–23 (2014)
30. Hardin, C.C., Taylor, A.D.: A peculiar connection between the axiom of choice and predicting the future. Am. Math. Mon. **115**, 91–96 (2008)
31. Hartogs, F.: Über das Problem der Wohlordnung. Mathematische Annalen **76**, 436–443 (1915)
32. Henkin, L.: The completeness of the first-order functional calculus. J. Symb. Log. **14**, 159–166 (1949)
33. Hilbert, D.: Mathematical problems. Bullet. AMS **8**(10), 437–479 (1902)
34. Hilbert, D.: Über das Unendliche. Mathematische Annalen **95**(1), 161–190 (1926); English translation in [18], pp. 367–392
35. Hodges, A.: Alan Turing: The Enigma. Random House, London (1992)
36. Hodges, W.: Läuchli's algebraic closure of $\mathbb{Q}$. Math. Proc. Camb. Philos. Soc. **2**, 289–297 (1976)
37. James, I.: Remarkable Mathematicians. Cambridge University Press, Washington, DC (2003)
38. Jech, T.J.: The Axiom of Choice. Studies in Logic and the Foundations of Mathematics, vol. 75. North-Holland (1973); Reprinted by Dover, New York (2008)
39. Kanamori, A.: The Higher Infinite. Springer Monographs in Mathematics. Springer, London (2009)
40. Kaye, R.: Models of Peano Arithmetic. Oxford Logic Guides, vol. 15. Oxford University Press, Oxford (1991)
41. Kechris, A.: Classical Descriptive Set Theory. Graduate Texts in Mathematics, vol. 156. Springer, New York (1995)
42. Kelley, J.L.: The Tychonoff product theorem implies the axiom of choice. Fundamenta Mathematicae **37**, 75–76 (1950)
43. Kohlenbach, U.: Applied Proof Theory: Proof Interpretations and Their Use in Mathematics. Springer Monographs in Mathematics, vol. XX. Springer, Berlin/Heidelberg (2008)
44. Kreisel, G.: Hilbert's programme. Dialectica **12**(3–4), 346–372 (1958)
45. Kreisel, G., Krivine, J.L.: Elements of Mathematical Logic. Studies in Logic. North-Holland, Amsterdam (1967)
46. Lambek, J., Scott, P.J.: Introduction to Higher Order Categorical Logic. Cambridge University Press, Cambridge (1986)
47. Lang, S.: Algebra. Graduate Texts in Mathematics, vol. 211. Springer, Princeton (2002)
48. Mac Lane, S., Moerdijk, I.: Sheaves in Geometry and Logic. Springer, New York (1992)
49. Marker, D.: Model Theory – An Introduction. Graduate Texts in Mathematics, vol. 217 Springer, Berlin (2002)

50. Marker, D.: Lectures on Infinitary Model Theory. Lecture Notes in Logic, vol. 46. Cambridge University Press, Cambridge (2016)
51. Marker, D., Messmer, M., Pillay, A.: Model Theory of Fields. Lecture Notes in Logic, vol. 5, 2nd edn. Association for Symbolic Logic (2002)
52. Matiyasevich, Y.: On Hilbert's Tenth Problem. Lecture Notes. University of Calgary (2000). Available at http://www.mathtube.org/sites/default/files/lecture-notes/Matiyasevich.pdf
53. Matyasevich, Y.: Hilbert's Tenth Problem. MIT Press, Cambridge (1993)
54. Menzler–Trott, E.: Logic's Lost Genius: The Life of Gerhard Gentzen. History of Mathematics, vol. 33. American Mathematical Society and London Mathematical Society, Providence (2007)
55. Munkres, J.R.: Topology, 2nd edn. Prentice Hall, Upper Saddle River (2000)
56. Odifreddi, P.: Classical Recursion Theory. Studies in Logic, vol. 125. North-Holland, Amsterdam (1989)
57. Odifreddi, P.: Classical Recursion Theory II. Studies in Logic, vol. 143. North-Holland, Amsterdam (1999)
58. Pillay, A.: An Introduction to Stability Theory. Oxford Logic Guides, vol. 8. Oxford University Press, New York (1983); Reprinted 2008 by Dover
59. Pillay, A.: Geometric Stability Theory. Oxford Logic Guides, vol. 32. Oxford University Press, New York (1996)
60. Pohlers, W.: Proof Theory. Universitext. Springer, Berlin/Heidelberg (2009)
61. Poonen, B.: Undecidability in number theory. Not. Am. Math. Soc. 55(3), 344–350 (2008)
62. Prawitz, D.: Natural Deduction – A Proof-Theoretical Study. Dover, Mineola (2006); Reprint of 1965 Ph.D. thesis
63. Rathjen, M.: The realm of ordinal analysis. In: Cooper, B., Truss, J. (eds.) Sets and Proofs, pp. 219–279. Cambridge University Press, Cambridge (1999)
64. Robinson, A.: Non-standard Analysis. North-Holland, Amsterdam (1966)
65. Rogers, H.: Theory of Recursive Functions and Effective Computability. McGraw-Hill, New York (1967); Reprinted by MIT Press, Cambridge (1987)
66. Rubin, H., Rubin, J.E.: Equivalents of the Axiom of Choice. Studies in Logic, vol. 116. North-Holland, New York (1985)
67. Sacks, G.E.: Saturated Model Theory. Mathematics Lecture Note Series. Benjamin Inc, Reading (1972)
68. Schwichtenberg, H., Wainer, S.S.: Proofs and Computations. Perspectives in Logic. Cambridge University Press, Cambridge (2012)
69. Sieg, W.: Hilbert's Programme and Beyond. Oxford University Press, Oxford (2013)
70. Simpson, S.G.: Subsystems of Second Order Arithmetic, 2nd edn. Cambridge University Press, New York (2009)
71. Smoryński, C.: Hilbert's programme. CWI Q. 1(4), 3–59 (1988)
72. Soare, R.I.: Recursively Enumerable Sets and Degrees. Perspectives in Mathematical Logic. Springer, Berlin (1987)
73. Solovay, R.M.: A model of set theory in which every set of reals is measurable. Ann. Math. Second Ser. 92, 1–56 (1970)
74. Specker, E.: Zur Axiomatik der Mengenlehre (Fundierungs- und Auswahlaxiom). Zeitschrift für mathematische Logik und Grundlagen der Mathematik 3(3), 173–210 (1957)
75. Tarski, A.: Der Wahrheitsbegriff in den formalisierten Sprachen. Studia Philosophica 1, 261–405 (1935)
76. Tent, K., Ziegler, M.: A Course in Model Theory. Lecture Notes in Logic, vol. 40. Cambridge University Press, New York (2012)
77. Troelstra, A.S., Schwichtenberg, H.: Basic Proof Theory. Cambridge Tracts in Theoretical Computer Science, vol. 43. Cambridge University Press, Cambridge (1996)
78. Turing, A.: On computable numbers, with an application to the Entscheidungsproblem. Proc. Lond. Math. Soc. Ser. 2 42, 230–265 (1936); Reprinted in [13]

79. The Univalent Foundations Program: Homotopy Type Theory. Institute for Advanced Study, Princeton (2013). Collectively written account of HoTT by participants of the Special Year on Univalent Foundations of Mathematics, organized by Awodey, S., Coquand, Th., Voevodsky, V
80. van den Dries, L.: Lectures on Motivic Integration. Available at https://faculty.math.illinois.edu/~vddries/mo.dvi
81. van den Dries, L.: Tame Topology and O-Minimal Structures. London Mathematical Society Lecture Note Series, vol. 248. Cambridge University Press, Cambridge (1998)
82. van Mill, J.: An introduction to $\beta\omega$. In: Kunen, K., Vaughan, J.E. (eds.) Handbook of Set-Theoretic Topology, pp. 503–568. North-Holland, Oxford (1984)
83. Wang, H.: Reflections on Kurt Gödel. MIT Press, Cambridge (1987)
84. Wiedijk, F.: Formal proof – getting started. Not. AMS **55**(11), 1408–1414 (2008)
85. Zermelo, E.: Beweis, daß jede Menge wohlgeordnet werden kann. Mathematische Annalen **59**, 514–516 (1904)
86. Zermelo, E.: Untersuchungen über die Grundlagen der Mengenlehre. Mathematische Annalen **65**, 261–281 (1908)
87. Zilber, B.: Zariski Geometries. London Mathematical Society Lecture Note Series, vol. 360. Cambridge University Press, Cambridge (2010)

Index

© Springer Nature Switzerland AG 2018
I. Moerdijk, J. van Oosten, *Sets, Models and Proofs*, Springer Undergraduate
Mathematics Series, https://doi.org/10.1007/978-3-319-92414-4

Printed in the United States
By Bookmasters